Ion Beams and
Nano-Engineering

MATERIALS RESEARCH SOCIETY
SYMPOSIUM PROCEEDINGS VOLUME 1181

Ion Beams and Nano-Engineering

Symposium held April 14–17, 2009, San Francisco, California, U.S.A.

EDITORS:

Daryush ILA
Alabama A&M University
Normal, Alabama, U.S.A.

Paul K. Chu
City University of Hong Kong
Kowloon, Hong Kong

Jörg K.N. Lindner
Universität Paderborn
Paderborn, Germany

Naoki Kishimoto
National Institute for Materials Science
Tsukuba, Japan

John E.E. Baglin
IBM Almaden Research Center
San Jose, California, U.S.A.

Materials Research Society
Warrendale, Pennsylvania

CAMBRIDGE UNIVERSITY PRESS
Cambridge, New York, Melbourne, Madrid, Cape Town,
Singapore, São Paulo, Delhi, Mexico City

Cambridge University Press
32 Avenue of the Americas, New York NY 10013-2473, USA

Published in the United States of America by Cambridge University Press, New York

www.cambridge.org
Information on this title: www.cambridge.org/9781107408234

Materials Research Society
506 Keystone Drive, Warrendale, PA 15086
http://www.mrs.org

First published 2010
First paperback edition 2012

Single article reprints from this publication are available through
University Microfilms Inc., 300 North Zeeb Road, Ann Arbor, MI 48106

CODEN: MRSPDH

ISBN 978-1-107-40823-4 Paperback

BEAM LITHOGRAPHY, PATTERN
FORMATION AND NANOWIRES

*Invited Paper

MAGNETIC, OPTICAL AND SEMICONDUCTOR APPLICATIONS

NANOSTRUCTURE FORMATION AND FABRICATION OF 3D STRUCTURES

*Invited Paper

ION-SOLID INTERACTIONS

*Invited Paper

*BIOLOGICAL AND BIOMEDICAL
APPLICATIONS*

*Invited Paper

PREFACE

This volume contains papers presented at Symposium DD, "Ion Beams and Nano-Engineering," held on April 14–17 at the 2009 MRS Spring Meeting in San Francisco, California.

Ion beam techniques provide unique capabilities for exploring and custom-tailoring the properties, structure, interactions, and configuration of polymeric materials, biomolecular materials, and bio-compatible materials. New understanding of ion beam-matter interactions, and new facilities for precision ion beam processing now enable applications ranging from nano-fabrication, nano-patterning and high resolution resist lithography, and nanoparticle self assembly, to selective activation of surfaces, manipulation of cells, fabrication of bio-compatible materials and fabrication of structures for 3D structures. Such facilities enable advanced development of coatings and structures displaying engineered mechanical, optical, electronic, magnetic and chemical properties. Ion beam characterization techniques also offer new options involving high spatial resolution, ppb elemental sensitivity, and sub-micro-beam imaging. The presentations during this symposium and manuscripts in this proceedings volume deal with current and emerging applications of ion beam techniques.

Finally, the organizers acknowledge that the success of Symposium DD depended critically on the full participation of excellent researchers and scientists as well as graduate students. The organizers are also extremely grateful to MRS for making this symposium possible through its dedicated work before, during and following the Meeting, and to the volunteer students who kindly assisted with audio and video during the sessions. We also thank NASA (USA), NIMS (Japan), AAMURI (USA), and National Electrostatics Corporation for their generous and substantial financial support.

Daryush ILA
Paul K. Chu
Jörg K.N. Lindner
Naoki Kishimoto
John E.E. Baglin

December 2009

MATERIALS RESEARCH SOCIETY SYMPOSIUM PROCEEDINGS

MATERIALS RESEARCH SOCIETY SYMPOSIUM PROCEEDINGS

Volume 1175E —Novel Functional Properties at Oxide-Oxide Interfaces, G. Rijnders, R. Pentcheva, J. Chakhalian, I. Bozovic, 2009, ISBN 978-1-60511-148-3

Volume 1176E —Nanocrystalline Materials as Precursors for Complex Multifunctional Structures through Chemical Transformations and Self Assembly, Y. Yin, Y. Sun, D. Talapin, H. Yang, 2009, ISBN 978-1-60511-149-0

Volume 1177E —Computational Nanoscience — How to Exploit Synergy between Predictive Simulations and Experiment, G. Galli, D. Johnson, M. Hybertsen, S. Shankar, 2009, ISBN 978-1-60511-150-6

Volume 1178E —Semiconductor Nanowires — Growth, Size-Dependent Properties and Applications, A. Javey, 2009, ISBN 978-1-60511-151-3

Volume 1179E —Material Systems and Processes for Three-Dimensional Micro- and Nanoscale Fabrication and Lithography, S.M. Kuebler, V.T. Milam, 2009, ISBN 978-1-60511-152-0

Volume 1180E —Nanoscale Functionalization and New Discoveries in Modern Superconductivity, R. Feenstra, D.C. Larbalestier, B. Maiorov, M. Putti, Y.-Y. Xie, 2009, ISBN 978-1-60511-153-7

Volume 1181 — Ion Beams and Nano-Engineering, D. Ila, P.K. Chu, N. Kishimoto, J.K.N. Lindner, J. Baglin, 2009, ISBN 978-1-60511-154-4

Volume 1182 — Materials for Nanophotonics — Plasmonics, Metamaterials and Light Localization, M. Brongersma, L. Dal Negro, J.M. Fukumoto, L. Novotny, 2009, ISBN 978-1-60511-155-1

Volume 1183 — Novel Materials and Devices for Spintronics, O.G. Heinonen, S. Sanvito, V.A. Dediu, N. Rizzo, 2009, ISBN 978-1-60511-156-8

Volume 1184 — Electron Crystallography for Materials Research and Quantitative Characterization of Nanostructured Materials, P. Moeck, S. Hovmöller, S. Nicolopoulos, S. Rouvimov, V. Petkov, M. Gateshki, P. Fraundorf, 2009, ISBN 978-1-60511-157-5

Volume 1185 — Probing Mechanics at Nanoscale Dimensions, N. Tamura, A. Minor, C. Murray, L. Friedman, 2009, ISBN 978-1-60511-158-2

Volume 1186E —Nanoscale Electromechanics and Piezoresponse Force Microcopy of Inorganic, Macromolecular and Biological Systems, S.V. Kalinin, A.N. Morozovska, N. Valanoor, W. Brownell, 2009, ISBN 978-1-60511-159-9

Volume 1187 — Structure-Property Relationships in Biomineralized and Biomimetic Composites, D. Kisailus, L. Estroff, W. Landis, P. Zavattieri, H.S. Gupta, 2009, ISBN 978-1-60511-160-5

Volume 1188 — Architectured Multifunctional Materials, Y. Brechet, J.D. Embury, P.R. Onck, 2009, ISBN 978-1-60511-161-2

Volume 1189E —Synthesis of Bioinspired Hierarchical Soft and Hybrid Materials, S. Yang, F. Meldrum, N. Kotov, C. Li, 2009, ISBN 978-1-60511-162-9

Volume 1190 — Active Polymers, K. Gall, T. Ikeda, P. Shastri, A. Lendlein, 2009, ISBN 978-1-60511-163-6

Volume 1191 — Materials and Strategies for Lab-on-a-Chip — Biological Analysis, Cell-Material Interfaces and Fluidic Assembly of Nanostructures, S. Murthy, H. Zeringue, S. Khan, V. Ugaz, 2009, ISBN 978-1-60511-164-3

Volume 1192E —Materials and Devices for Flexible and Stretchable Electronics, S. Bauer, S.P. Lacour, T. Li, T. Someya, 2009, ISBN 978-1-60511-165-0

Volume 1193 — Scientific Basis for Nuclear Waste Management XXXIII, B.E. Burakov, A.S. Aloy, 2009, ISBN 978-1-60511-166-7

Prior Materials Research Society Symposium Proceedings available by contacting Materials Research Society

Beam Lithography, Pattern Formation, and Nanowires

Mater. Res. Soc. Symp. Proc. Vol. 1181 © 2009 Materials Research Society   1181-DD01-04

Localized ^{56}Fe$^+$ Ion Implantation of TiO$_2$ Using Anodic Porous Alumina

J. Jensen[1], R. Sanz[2], M. Jaafar[2], M. Hernández-Vélez[2,3], A. Asenjo[2], A. Hallén[4], M. Vázquez[1]

[1] Thin Film Physics Division, IFM, Linköping University, SE-581 83 Linköping, Sweden
[2] Instituto de Ciencia de Materiales de Madrid, CSIC, 28043 Madrid, Spain
[3] Applied Physics Department, Universidad Autónoma de Madrid, 28043 Madrid, Spain
[4] ICT-MAP, Royal Institute of Technology, SE-164 40 Stockholm, Sweden

ABSTRACT

We present result following localized ion implantation of rutile titanium dioxide (TiO$_2$) using anodic porous alumina as a mask. The implantation were performed with 100 keV ^{56}Fe$^+$ ions using a fluence of $1.3 \cdot 10^{16}$ ions/cm^2. The surface modifications where studied by means of SEM, AFM/MFM and XRD. A well-defined hexagonal pattern of modified material in the near surface structure is observed. Local examination of the implanted areas revealed no clear magnetic signal. However, a variation in mechanical and electrostatic behavior between implanted and non-implanted zones is inferred from the variation in AFM signals.

INTRODUCTION

Regular nano- and micro-patterns have potential technological application in *e.g.* data storage, biological sensors or photonic crystals. Ion implantation is an ideal instrument to modify material properties in a controlled way [1], and very suitable for fabrication and tailoring of functional properties such as optical/magnetic patterns, nanoporous material and catalyst surfaces. An important issue is to control the specific implantation areas. The spatial controls of areas where the material changes take place are fundamental in order to develop functional devices or study collective responses. One way of creating regular structures, or an array of modulated physical properties, is with focused ion beam implantation (FIB). However, since currently only a few ion sources and restricted energies are available, the application of this method is limited. Another method is using ion beam-based projection methods, where the ion impact is restricted by a mask or template, enabling a parallel formation of well-ordered and localized material modification yielding structures with well-defined features and interesting material properties [2-4]. A special type of implantation mask, which within the last few years has received much attention, is anodic porous alumina membranes, which contain self-ordered nanopores. Until now only few studies have been performed with this mask for ion beam nanolithography using keV ions [5,6,7].

Titanium dioxide (TiO$_2$) is a particularly versatile material with extensive technological application. Implanting this material with ferromagnetic ions makes it possible to obtain a diluted magnetic semiconductor [8], to be applied in future spintronics devices. Here one wants to avoid the clustering of doped species. However, for other application like magnetooptics, magneto-transport, and nanomagnetism the system must be composed of nanoclusters dispersed in a semiconductor or dielectric matrices [1]. Previous results on continuum implanted TiO$_2$ using 100 keV Fe ions [9] with fluences from $2 \cdot 10^{16}$ to $1 \cdot 10^{17}$ cm^2 showed the precipitation of metastable compounds such as FeTi$_2$O$_5$. More recently 180 keV Fe implanted rutile TiO$_2$ [10] was studied, presenting the precipitation and thermal evolution of epitaxial Fe and FeTiO$_3$

nanoclusters paying attention to their magnetic properties. In addition to interesting magnetic properties, doped TiO_2 is an extensively studied photo-catalytic semiconductor [11]. Doping of TiO_2 with Fe has been shown to enhance the efficiency of its photoactivity [12]. Even the optical properties of TiO_2 can be modified by ion irradiation [13].

In this work we present first results following localized ion implantation of rutile TiO_2 single crystal using 100 keV $^{56}Fe^+$ ions. The localized ion implantation was accomplished by irradiating the substrate through an anodic porous alumina membrane.

EXPERIMENTAL

Rutile single crystals (MTI Coorporation) with a <110> surface orientation were used as substrates (ρ=4.25 g cm^{-3}). The employed mask where 2 μm thick anodic porous alumina membrane (PAM) with 265 nm diameter and 450 nm lattice parameter, see Figure 1(a), obtained by a double anodization methods described elsewhere [14]. Prior to implantation the mask were placed on the substrate surface.

The samples were implanted with 100 keV $^{56}Fe^+$ ions at room temperature to a total fluence of 1.3•10^{16} ions/cm^2. The average scanned ion beam flux was kept constant at about 7·10^{11} cm^{-2} s^{-1}. Simulations of the implantation profiles was carried out by the TRIM (version SRIM2008) code [15], suggesting a near Gaussian ion distribution with a projected range of 50 nm and a straggling of 20 nm.

As the aspect ratio of the pores in the mask is small, giving an acceptance angle of approximately 5°, no alignment procedure was employed [16] and the irradiation was thus done at normal incidence. According to the TRIM code, the 100 keV Fe ions are completely stopped by the 2 μm thick PAM material (ρ= 2.3 g cm^{-3} [16]), so substrate areas between the pores should not be exposed directly to the ion beam. The lateral projected range and straggling is estimated much smaller than the average pore distance, so there should be no overlap of implanted volumes and essential no damage between the pores. Consequently, the exposed areas of the substrates under the masks roughly correspond to 54 percent of the total area. The maximum peak concentration of Fe located within the implanted volumes is ≈3.0%, given by TRIM simulation.

After implantation the PAM was removed by immersing in deonized water followed by a cleaning process with acetone using an ultrasonic bath. A LEO 1550 FEG high-resolution scanning electron microscope (SEM) was used to investigate the surface morphology. The samples implanted through PAM were furthermore investigated by atomic force microscopy (AFM) and magnetic force microscopy (MFM). The AFM/MFM measurements were carried out in dynamic mode using a Cervantes model from Nanotec. We used commercial Nanosensor PPP-MFMR probes. Finally, structural analysis of the substrate was performed by means of grazing incidence (3°) X-ray diffraction measurements using Cu, K$_{\alpha1}$ (1.5604 Å) radiation and elemental analysis was performed by energy dispersive X-ray spectroscopy (EDS).

RESULTS AND DISCUSSION

The SEM observations of the irradiated sample show a clear contrast between implanted (damaged) and non-implanted areas, see Figure 1(b). The image shows an array of implanted zones in a hexagonal arrangement formed through the 'patterned' ion implantation. The sizes of the implanted zones are similar to the mask pore holes with no appreciable change in the shape or size of implanted areas compared to original mask features. The ion implantation through the PAM thus yields a replication of the mask features and well-defined pattern to the substrate surface using the present ion energy and fluence. Note the interesting contrast seen at the border between implanted and virgin surface area.

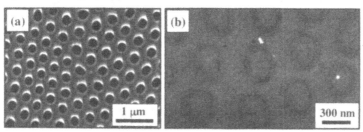

Figure 1. Scanning electron micrographs of (a) an anodic porous alumina membrane (PAM), the employed mask, and (b) the <110> rutile TiO_2 surface after ion implantation through such a mask with 100 keV $^{56}Fe^+$ to a fluence of $1.3 \cdot 10^{16}$ ions/cm^2. The mask was removed and the substrate cleaned before observation.

In order to check the existence of contaminations from the mask on the surface, EDS measurements were carried out. The EDS measurements showed an atomic composition of Ti and O in stoichiometric proportions. Even though Fe is detected, the total amount could not be exactly determined. The implanted Fe leads only to a small characteristic X-ray peak, which is residing on a large continuum background in the X-ray spectrum from the substrate. No traces from atomic species present in the PAM (Al and P) were detected, within the detection limit. Interesting is to note that even in an optical microscope a regular variation in optical reflectance was seen. This indicates a localized modification of the optical properties. Changes in optical absorbance of rutile have been observed after implantation with 64 keV Ni^+ ions [17].

To investigate the change in surface morphology after 'patterned' ion implantation, AFM measurements were performed, see Figure 2. A contrast, seen as a depth variation, is clearly observed. Non-implanted areas did not present any remarkable change of the surface morphology, having a maximum RMS roughness value of 0.69 nm. Implanted zones exhibit a crater shape with a convex structure inside with an average height of 2.5 nm from the bottom of the crater. A detailed profile of one of the structures is presented in Figure 2(b). The contrast seen by SEM in the implanted areas in Figure 1(b) is believed to be related to the depth profile observed by the AFM measurements.

The depression in the structures may originate by a combined effect of sputter erosion and a change in density caused by the induced ion damage. A simple TRIM simulation (for amorphous TiO_2) yields a sputter depth up to 6 nm for the applied fluence. However, the convex structure obtained in samples irradiated through the PAM, can be ascribed to the restricted implantation where other effects such as redeposition of sputtered material could be more noticeable than in not masked conventional implantations. Also, sputtering of mask material is possible although no trace of deposited mask material were detected.

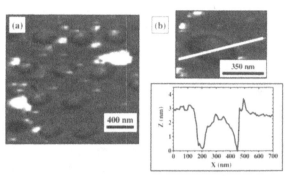

Figure 2. (a) AFM image of the rutile surface after implantations through a PAM mask with 100 keV Fe ions to fluence of $1.3 \cdot 10^{16}$ ions/cm^2. (b) The profile of an AFM line scan through an exposed area. The mask was removed and the substrate cleaned before observation.

The local magnetic behavior of the localized implanted samples was measured by MFM. In Figure 3 we present the topographic (a) and frequency shift images (b and c). Image contrast associated to a frequency shift, measured at the same tip-sample distance as the topography, resulted in a maximum value of 4 Hz. The origin of this variation can be ascribed to electrostatic interaction or changes in the mechanical properties such as visco-elasticity, adhesion and/or friction of the implanted areas. The contrast may even arise due to changes in dielectric properties [18]. This phenomenon of frequency shift has not been studied in detail at the present moment. But a local change in dielectric property may be explained as a combination of a change in the active surface chemistry of TiO$_2$ [19] and the effect of the defects originated by ion implantation. Another source for changes in the dielectric property is the existence of small Fe metal clusters surrounded by a dielectric material. To study the magnetic signal a second pass, where the tip- sample distance was increased to about 25 nm, was performed. We did not see any clear magnetic contrast from implanted and non-implanted areas, Figure 3(c). Observation of no magnetic contrast can originate by the low magnetic moment in the implanted zones, being lower than the detection threshold of the instrument.

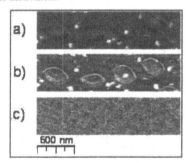

Figure 3. AFM and MFM images showing (a) topographic signal, (b) frequency shift, and (c) frequency shift at 25 nm from the surface of a rutile substrate after implantation with 100 keV ^{56}Fe$^+$ ions to a fluence of 1.3×10^{16} ions/cm^2 through a PAM.

Preliminary grazing incidence XRD measurements of the rutile substrate implanted with 100 keV $^{56}Fe^+$ ions, without the use of a mask (i.e. continuum implantation), is shown in Figure 4. The peaks seen in the virgin substrate are gone. However, in addition to an amorphous halo, broad peaks belonging to new phases are observed. The peaks are situated close to the positions of diverse Fe oxides and Fe-Ti compounds. However, taking into account the reported existence of metastable Fe-Ti precipitates generated by ion implantation processes [9] it is not possible to precisely assign the detected peaks by just relying on the XRD technique. The existence of metastable precipitates may justify the observed changes in dielectric properties. It can also explain the missing magnetic contrast, as seen by MFM, due to a superparamagnetism behavior of those precipitates or just a low magnetic moment.

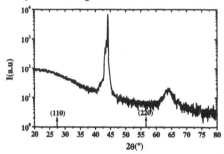

Figure 4. Grazing incidence XRD diffractogram of a <110> rutile substrate after continuum implantation with 100 keV $^{56}Fe^+$ ions to a fluence of $1.3x10^{16}$ ions/cm^2. The angle of incidence for GIXRD was 3°. The original rutile diffractions peak positions (110) and (220) are indicated with arrows.

In a recent study, where we implanted ZnO with 35 keV $^{55}Mn^+$ ions, we saw different results in samples implanted through PAM as compared to samples implanted over the whole surface (continuum implantations) [20]. The differences manifested themselves by *e.g.* dissimilar XRD spectra and magnetic response as observed by SQUID. In addition, it has previously been observed that localized ion implantation within micrometer size areas may lead to different diffusion and clustering properties as compare to continuum implantation as observed after Au ion implantation into TiO_2 [21,22]. It is thus suggestive that also in the present case of sub-micrometer restricted implantation, there might be differences in continuum and masked implantation and that the XRD data shown in Figure 4 are not representative for the masked implantation.

CONCLUSIONS AND OUTLOOK

Applying a PAM as stencil mask, localized zones have been successfully implanted, with an acceptable accuracy of pattern reproduction. The present stencil mask may therefore be used for creating well defined material contrast employing ion beam implantation techniques. The detected changes in local dielectric properties in the restricted implanted samples deserve a more carefully study due to the interesting applications of the active surface chemistry of TiO_2 and its optical properties.

There was no detectable magnetic response of restricted implanted zones as observed by MFM. However, it is not possible to rule out a ferromagnetic response referring to reported

7

results obtained under continuum ion implantation conditions. Taking into account the possibility for an existence weak ferromagnetic signal, other techniques, *e.g.* X-ray magnetic circular dichroism, should be employed to study this possibility in detail. These studies are currently in progress.

Nonetheless, localized ion implantation may lead to different diffusion and clustering properties as compare to continuum implantation, which need to be look into. This may have interesting consequences for magnetic, optical and surface chemical surface properties when performing restricted ion implantation.

ACKNOWLEDGMENTS

This work was supported in part by the Spanish Ministry of Education under Grant MAT2007-6042. J. Jensen thanks the Carl Tryggers Foundation and Linköping University through the VR Linneaus grant LiLi-NFM for financial support. M.J. Thanks the CAM for the financial support.

REFERENCES

1. A. Meldrum, R. F. Haglund, Jr., L. A. Boatner, and C. W. White. Adv. Mater. **13**, 1431 (2001).
2. T. Shibata, K. Suguro, K. Sugihara; T. Nishishashi; J. Fujiyama; Y. Sakurada, IEEE transactions on semiconductor manufacturing **15**, 183, (2002).
3. E. Knystautas, *'Engineering Thin Films and Nanostructures with Ion Beams'*, Optical Science and Engineering Series Vol. **95** CRC Press (2005).
4. *Ion-beam-based Nanofabrication*, edited by D. Ila, J. Baglin, N. Kishimoto, P.K. Chu, MRS symposium proc. Vol **1020** (2007), and MRS spring meeting 2009, symposium DD.
5. N. Matsuura *et al.*, Appl. Phys. Lett. **81**, 4826 (2002).
6. S.W. Shin *et al.*, Nanotechnology **16**, 1392 (2005).
7. M. Nakamura, S. Nigo, N. Kishimoto, Trans. Mater. Res. Soc. Jpn. **33**, 1101 (2008).
8. R. Janisch, P. Gopal, and N. A Spaldin. J. Phys.: Condens. Matter **17**, R657 (2005).
9. M. Guermazi *et al.*, Mat. Res. Bull. **18**, 529 (1983)
10. M. Zhou *et al.*, J. Appl. Phys. **103**, 083907 (2008).
11. A. Fujishima, K. Hashimoto, and T. Watanabe, *TiO₂ Photocatalysis: Fundaments and Applications*, BKC, Tokio, (1999).
12. C. Adán, A. Bahamonde, M. Fernández-García, A. Martínez-Arias, Appl. Catal. B **72**, 11 (2007).
13. J. Jensen, M. Skupinski, K. Hjort, R. Sanz, Nucl. Instrum. and Methods B **266**, 3113 (2008).
14. H. Masuda and F. Fukuda, Science **268**, 1466 (1995).
15. www.srim.org.
16. R. Sanz, J. Jensen, A. Johansson, M. Skupinski, G. Possnert, M. Boman, M. Hernandez-Vélez, M. Vazquez, K. Hjort. Nanotechnology **18**, 305303 (2007).
17. S. Zhu, L.M. Wang, X.T. Zu, and X. Xiang, Appl. Phys. Lett. **88**, 043107 (2006).
18. C. Dumas *et al.*, Microelectronic Engineering **85**, 2358 (2008).
19. U. Diebold, Surf. Sci. Rep. **48**, 53 (2003).
20. R. Sanz, J. Jensen, G. González-Díaz, O. Martínez, M. Vázquez and M. Hernández-Vélez, Nanoscale Research. Lett., Accepted (2009).
21. K. Sun, S. Zhu, R. Fromknecht, G. Linker, L.M. Wang, Materials Letters **58**, 547 (2004).
22. R. Fromknecht, G. Linker, L.M. Wang, S. Zhu, K. Sun, A. van Veen, M. van Huis, J. Niemeyer, T. Weimann, J. Wang, Surf. Interface Anal. **36**, 193 (2004).

8

Mater. Res. Soc. Symp. Proc. Vol. 1181 © 2009 Materials Research Society   1181-DD13-13

Importance of internal ion beam parameters on the self-organized pattern formation with low-energy broad beam ion sources

Marina I. Cornejo, Bashkim Ziberi, Michael Tartz, Horst Neumann, Frank Frost, Bernd Rauschenbach

Leibniz-Institut für Oberflächenmodifizierung e.V. (IOM) Permoserstr. 15, D-04318 Leipzig, Germany

E-mail: marina.cornejo@iom-leipzig.de

ABSTRACT

A first qualitative approach to the importance of the divergence angle and angular distribution of the ions within the broad beam (here called internal beam parameters) on the pattern formation by low-energy ion beam erosion is presented. Si (100) surfaces were irradiated with Kr^+, with an ion energy of 2 keV, using a Kaufman type broad beam ion source. It is found that the operating parameters of the broad beam ion source which are responsible for the angular distribution of the ions also affect the pattern formation. Especially, the effect of the acceleration voltage, discharge voltage, grid distance and operation time on the transition from ripple to dot pattern with increasing ion beam incidence angle were analyzed. The results represent additional evidence about the significance of the internal beam parameters and the need of the further investigation of their role on the pattern formation by low-energy erosion.

INTRODUCTION

The low-energy noble gas ion beam erosion of solid surfaces is a simple bottom-up approach for the generation of nanostructures. For certain sputtering conditions well ordered self-organized nanostructures (e.g., ripples, dots) can be formed. Due to the use of broad beam ion sources, low-energy ion beam erosion is particularly suitable as a cost-efficient method to produce large-area nanostructured surfaces in a one-step process.

The surface topography evolution is, in general, attributed to the competition of curvature dependant sputtering that roughens the surface and smoothing by different surface relaxation mechanisms [1-5]. It is also well known that the incidence angle of the ions is a critical parameter that determines the surface topography. On Si surfaces, different topographies emerge on the surface due to the bombardment (without rotation of the sample) at different ion beam incidence angles [6]. Ion beam incidence angle refers here to the angle between the substrate normal and the ion source axis (geometrically defined incidence angle). Inherent to all broad beam ion sources, which are essential for large area processing and often used for low-energy ion beam erosion, the ion beam exhibits a certain divergence, i.e. the ion trajectories are not strictly parallel to each other. This generates a spread of the local incidence angles with respect to the geometrically defined ion beam incidence angle. There are only few works about the patterning by low-energy ion beam erosion where the angular distribution was contemplated and considered important for the surface topography evolution [7, 8]. There could be a connection between the variety of results obtained by the different research groups [9-12] and the divergence angle and angular distribution of the ions, here called internal ion beam parameters.

In this work the effect of some operational parameters of a broad beam Kaufman type ion source on the surface topography evolution is presented. Among all the experimental parameters that affect the pattern formation by low-energy ion beam erosion here the focus was set on the influence of those that affect the internal ion beam parameters. First, the effect of the accelerator voltage, i.e. the voltage applied to the accelerator grid is shown. Additionally, the importance of the discharge voltage, valid for Kaufman type broad beam ion sources, was analyzed. The geometry of the extraction system is also known to affect the ion angular distribution. In this regard the effect of the distance between the extraction grids and the effect of the grid alteration with the time on the surface topography are presented.

EXPERIMENTAL DETAILS

The ion source used for the experiments is a home-built Kaufman-type broad beam ion source (beam diameter 180 mm) equipped with a two-grid multi-aperture ion optic system. The first grid (screen grid) is at floating potential. The positive beam potential is applied to the plasma anode. The second grid (accelerator grid) is at a negative potential with respect to the grounded chamber and sample stage. The total voltage of the beam is the sum of the voltages applied to both grids. The overall ion beam parameters as the divergence angle and the angular distribution of the ions within the beam are determined by the full set of ion source parameters: grid voltages, plasma parameters and their spatial distribution and grid geometry [13, 14]. The energy of the ions on the target, however, depends only on the potential difference between the plasma and the target (usually at ground) and is defined here by the beam voltage. More details about the broad beam ion source used are given elsewhere [15].

The samples used were commercially available epi-polished (100) Si substrate (p type and 0.01 and 0.02 Ω cm) with a root-mean-square (rms) roughness of ~0.2 nm. The samples were mounted on a water cooling (temperature approx. 285 K) substrate holder in a high vacuum chamber with a base pressure of 1×10^{-6} mbar. The ion beam incidence angle (α_{ion}) can be varied from 0° and 90° in steps of five degrees. An additional angle adjustment is achieved using a sample holder with facets of one degree. The samples were irradiated with Kr^+ ions with a kinetic energy (E_{ion}) of 2 keV and a fluence (Φ) of 6.7×10^{18} cm^{-2}. The acceleration voltage (U_{acc}) and discharge voltage (U_{dis}) were varied from -200 to -1000 V and 50 to 150 V respectively. The distance between the two grids (d) was changed from 1 mm to 2 mm in some of the experiments in order to analyze the influence of this parameter. The surface topography was analyzed by scanning force microscopy (AFM) using a Dimension 3000 system with Nanoscope IIIa controller from Digital Instruments operating in TappingMode^{TM}. The AFM results presented have a resolution of 512 × 512 pixels, and Si probes with (nominal) tip radii smaller than 10 nm were used.

RESULTS AND DISCUSSION

In Fig. 1 AFM images of different topographies emerging on Si surfaces when irradiated at different ion beam incidence angles without sample rotation are presented. It is shown that increasing the ion beam incidence angle from 15 to 30°, the transition from ripples to smooth surface takes place. An intermediate transition structure (here called transition structure ripples-dots) is clearly observed (Fig. 1b). The ripples begin to transform into dots ordered along the previously existing ripples, as shown here for E_{ion}= 2 keV and U_{acc}= -1000 V.

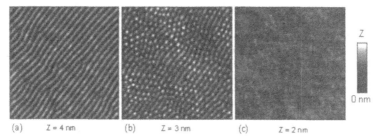

Figure 1. 1 μm × 1 μm AFM images of Si surfaces after irradiation with Kr$^+$, E_{ion} = 2 keV, Φ = 6.7 × 10^{18} cm^{-2}, U_{acc} = - 1000 V, U_{dis} = 100 V, d = 1 mm and α_{ion} = (a) 15°, (b) 24° and (c) 30°.

Applying an acceleration voltage of -600 V, the transition takes place at lower ion beam incidence angles. There is a shift to lower angles. The results are summarized in Fig. 2a showing the topography diagram for Si for different ion beam incidence angles and acceleration voltages. It is seen from this diagram that for U_{acc} = -200 V no pattern evolves on the surface.

The acceleration voltage affects the ion angular distribution (but not the ion energy). In a previous work [7] this effect was studied by ion beam simulations. This study was done for the same broad beam ion source used here and similar operating conditions. The simulations showed that increasing the accelerator voltage the angular distribution broadens and the maximum shifts towards larger angles. Therefore, the spread of ions arriving at the target becomes larger with the accelerator voltage. According to the results shown here in the topography diagram (Fig. 2a) the formation of ripples seems to be related with larger angular distribution of the ions.

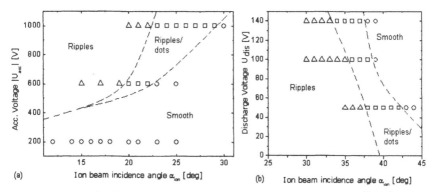

Figure 2. Topography diagrams for Si surfaces for E_{ion}= 2 keV and Φ = 6.7 × 10^{18} cm^{-2}, d = 1 mm △: ripples, ◯: smooth surface, ☐: transition ripples-dots. The dash lines are only a guide to the eye.
(a) for different acceleration voltages U_{acc} and ion beam incidence angles α_{ion} (U_{dis} = 100 V)
(b) for different discharge voltages U_{dis} and ion beam incidence angle α_{ion} (U_{acc} = - 1000 V)

The shape and position of the plasma sheath determine also the angular distribution. The plasma sheath is expected to vary in position and shape as a result of plasma properties, accelerator system geometry and accelerator system potential variations. The discharge voltage controls the acceleration of the emitted electrons in the filament sheath, which in turn affects the plasma sheath shape and position [14] and in consequence also the angular distribution of the ions within the beam. The way the discharge voltage affects the internal beam parameters is not easily identified. Together with the discharge voltage, the influence of the plasma properties and the ion optics system geometry should be contemplated. A further effect of the increasing discharge voltage is the increasing content of double charged ions in the plasma and the beam. In the topography diagram for different discharge voltages and ion beam incidence angles (Fig. 2b) it is observed that the transition from ripples to dots takes place at different angles depending on the discharge voltage. The transition angle decreases when higher discharge voltages are applied.

The geometry of the extraction system also affects the internal beam parameters [13]. In this regard Si samples were irradiated using two different distances between the grids. It was observed that the change of the grid distance leads to a shift on the angle at which the transition from ripple to dot pattern takes place. For distances of 1 mm and 2 mm the ripples are stable up to an ion beam incidence angle of 36° and 26° respectively. Figure 3 shows simulated beamlets (ion beam extracted by one grid hole) using the IGUN code [16] for a grid distance of 1 mm (left) and 2 mm (right) at various plasma densities. At lower plasma density (upper plots) the increase of the grid distance decreases the beamlet divergence due to the reduced focusing strength of the grid potential distribution. At larger plasma densities (bottom plots), however, the larger grid distance leads to a higher beamlet divergence. It can be seen that the change of the grid distance changes the angular ion distribution and, consequently, may affect the pattern formation too. However, the effect of the grid distance on the beam properties depends on the actual plasma parameters, further experimental investigations are necessary.

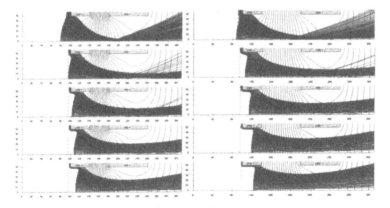

Figure 3. Simulated beamlets at various plasma densities and grid distances (left: 1 mm, right: 2 mm) for $E_{ion} = 2$ keV, $U_{acc} = -1000$ V, $U_{dis} = 100$ V. The plasma density from top to bottom: 5×10^{10} cm^{-3}, 7.5×10^{10}, 1×10^{11}, 2.5×10^{11} and 5×10^{11} cm^{-3}.

In addition, an effect of the ion source operation time on the topography evolution could be observed. It is known that due to the collisions of the ions with the extraction grid alterations take place. Due to the grid erosion, the hole diameter increases with the operation time. Tartz et al. presented a grid erosion code developed to predict the grid alteration and the validation of the results with experimental data [17]. In regard to the ripple to dot pattern transition, the evolving topography on Si surfaces using a new grid system and after 250 hours of use was compared and a shift of four degrees on the ion beam incidence angle at which the transition takes place was observed. This could be related again to a change of the internal beam parameters due to the grid erosion. The increase of the hole diameter in the accelerator grid reduces the beamlet divergence. Experimentally it was observed that with the time the range of angles at which the ripples are stable is reduced. Ripples are stable up to 26° for the new grid system and up to 22° after 250 hs operation. These results seem to agree with the proposition that a certain ion angular distribution is necessary for the ripples stability.

CONCLUSIONS

In this contribution a first approach to the importance of the operating parameters that determine the internal ion beam parameters (divergence angle and angular distribution) on the pattern formation by low-energy ion beam erosion was presented. The divergence and angular distribution of the ions within the beam are inherent to all broad beam ion sources. As broad beam ion sources are essential for technological applications, it is necessary to achieve a better understanding of the role of these parameters for the reliability and reproducibility of the patterning. Also the comprehension of the importance of the internal beam parameters would represent a significant contribution for the understanding of the processes involved on the pattern formation. In this study the effect of the acceleration voltage, discharge voltage, grid distance and operation time on the pattern transition from ripples to dots on Si surfaces irradiated with 2000 eV Kr^+ ions have been shown. The acceleration voltage alters the ion beam incidence angle at which the pattern transition takes place. Such effect is thought to be related to a change on the angular distribution of the ions, i.e. on the local incidence angle. When the ions leave the extraction system with relatively small angular distributions, no pattern seems to evolve on the irradiated surface. It was also shown that the discharge voltage affects the angle range at which the transition occurs. This effect is considered to be connected to changes in the plasma sheath shape and position by alteration of the plasma properties and potential of the extraction system. The angular distribution is also partially determined by the grid system geometry and here it was shown that the grid distance affects strongly the pattern transition. Finally it was observed that the time of operation alters the transition, probably also due to changes of the grid geometry by ion erosion.

In the present, work is going on to establish a simple technique to determine the angular distribution in order to obtain a quantitative approach of its effect on the topography evolution. Also the connection between the angular distribution, potential substrate contamination (in particular by Fe) and the pattern formation is being analyzed.

ACKNOWLEDGEMENTS

This work is supported by the Deutsche Forschungsgemeinschaft (FOR845)

REFERENCES

[1] G. Carter, "The physics and applications of ion beam erosion," *Journal of Physics D: Applied Physics*, vol. 34, p. 22, 2001.
[2] W. L. Chan and E. Chason, "Making waves: Kinetic processes controlling surface evolution during low energy ion sputtering," *Journal of Applied Physics*, vol. 101, p. 46, 2007.
[3] R. Cuerno, H. A. Makse, S. Tomassone, S. T. Harrington, and H. E. Stanley, "Stochastic Model for Surface Erosion via Ion Sputtering: Dynamical Evolution from Ripple Morphology to Rough Morphology," *Physical Review Letters*, vol. 75, p. 4, 1995.
[4] M. A. Makeev, R. Cuerno, and A.-L. Barabási, "Morphology of ion-sputtered surfaces," *Nuclear Instruments & Methods in Physics Research, section B*, vol. 197, p. 43, 2002.
[5] U. Valbusa, C. Boragno, and F. Buatier de Mongeot, "Nanostructuring surfaces by ion sputtering," *J. Phys.: Condens. Matter*, vol. 14, p. 22, 2002.
[6] B. Ziberi, "Ion Beam Induced Pattern Formation on Si and Ge Surfaces," in *Fakultät für Physik und Geowissenschaften*. vol. Doktor Leipzig: Leipzig University, 2006.
[7] B. Ziberi, F. Frost, H. Neumann, and B. Rauschenbach, "Ripple rotation, pattern transitions, and long range ordered dots on silicon by ion beam erosion," *Applied Physics Letters*, vol. 92, p. 063102, 2008.
[8] B. Ziberi, F. Frost, M. Tartz, H. Neumann, and B. Rauschenbach, "Importance of ion beam parameters on self-organized pattern formation on semiconductor surfaces by ion beam erosion," *Thin Solid Films*, vol. 459, pp. 106-110, 2004.
[9] D. Carbone, A. Alija, O. Plantevin, R. Gago, S. Facsko, and T. H. Metzger, "Early stage of ripple formation on Ge(001) surfaces under near-normal ion beam sputtering," *Nanotechnology*, vol. 19, p. 5, 2008.
[10] A. Cuenat and M. J. Aziz, "Spontaneous Pattern Formation from Focused and Unfocused Ion Beam Irradiation," *Materials Research Society Symposia Proceedings*, vol. 969, p. 6, 2002.
[11] C. Hofer, S. Abermann, C. Teichert, T. Bobek, H. Kurz, K. Lyutovich, and E. Kasper, "Ion bombardment induced morphology modifications on self-organized semiconductors surfaces," *Nuclear Instruments & Methods in Physics Research, section b*, vol. 216, p. 7, 2004.
[12] F. J. Ludwig, C. R. J. Eddy, O. Malis, and R. L. Headrick, "Si(100) surface morphology evolution during normal-incidence sputtering with 100-500 eV Ar$^+$ ions," *Applied Physics Letters*, vol. 81, p. 3, 2002.
[13] M. Tartz, "Simulation des Ladungstransportes in Breitstrahlionenquellen " in *Fakultät für Physik und Geowissenschaften*. vol. Doktor Leipzig: Leipzig University, 2003.
[14] M. Zeuner, H. Neumann, F. Scholze, D. Flamm, M. Tartz, and F. Bigl, "Characterization of a modular broad beam ion source," *Plasma Sources Sci. Technology*, vol. 7, pp. 252-267, 1998.
[15] F. Frost, B. Ziberi, A. Schindler, and B. Rauschenbach, "Surface engineering with ion beams: from self-organized nanostructures to ultra-smooth surfaces," *Appl. Phys. A*, vol. 91, p. 9, 2008.
[16] R. Becker and W. B. Herrmannsfeldt, "IGUN- A program for the simulation of positive ion extraction including magnetic fields," *Review of Scientific Instruments*, vol. 63, p. 3, 1992.
[17] M. Tartz, E. Hartmann, and H. Neumann, "Validated simulation for the ion extraction grid lifetime," *Review of Scientific Instruments*, vol. 79, p. 02B905, 2008.

Mater. Res. Soc. Symp. Proc. Vol. 1181 © 2009 Materials Research Society   1181-DD10-02

Nanoscale Surface Patterning of Silicon Using Local Swelling Induced by He Implantation Through NSL-Masks

Frederic J.C. Fischer[1], Michael Weinl[1], Jörg K.N. Lindner[1,2], and Bernd Stritzker[1]
[1]Universität Augsburg, Institut für Physik, D-86135 Augsburg, Germany
[2] present address: Universität Paderborn, Department Physik, D-33098 Paderborn, Germany

ABSTRACT

A novel technique to form periodically nanostructured Si surface morphologies based on nanosphere lithography (NSL) and He ion implantation induced swelling is studied in detail. It is shown that by implantation of keV He ions through the nanometric openings of NSL masks regular arrays of hillocks and rings can be created on silicon surfaces. The shape and size of these surface features can be easily controlled by adjusting the ion dose and energy as well as the mask size. Feature heights of more than 100 nm can be obtained, while feature distances are typically 1.15 or 2 (hillock or ring) nanosphere radii, which are chosen to be between 100 and 500 nm in this study. Atomic force and scanning electron microscopy measurements of the surface morphology are supplemented by cross-sectional transmission electron microscopy, revealing the inner structure of hillocks to consist of a central cavity surrounded by a hierarchical arrangement of smaller voids. The surface morphologies developed here have the potential to be useful for fixing and separating nano-objects on a silicon surface.

INTRODUCTION

Nanosphere Lithography (NSL) is a cheap and versatile bottom-up method to create regular arrays of nanoscale surface features [1]. Usually, for this purpose equally-sized nanospheres from a colloidal suspension are self-assembled in a hexagonally close-packed mono- or double layer on a substrate and the free space between each triple of neighbouring spheres is used as a mask opening in a subsequent materials deposition process. In the present paper, NSL masks are used as ion implantation masks. While NSL combined with physical vapour deposition is being used by many groups, the implantation of keV ions through NSL masks has only recently attracted some attention [2,3]. It has been shown that NSL masks made of either SiO_2 or polystyrene (PS) beads with diameters of few hundrets of nanometers are comparatively stable during ion bombardment up to medium doses while at high doses deformations occur due to ion beam induced sintering at the sphere contacts and other effects.

Here we report on the patterned high-dose implantation of keV He ions through NSL masks into silicon. He implantation into bare Si has been extensively studied in the past due to the possibility of layer splitting and impurity gettering in voids [4,5]. As we have shown in a recent study [6] such implantations through NSL masks lead to a localized swelling of the silicon underneath the mask openings, resulting in a regular surface pattern of hillocks. The aim of the present paper is to study the dependence of the hillock height on the ion energy, the width of the mask openings and the dose, in order to allow for a precise tailoring of the surface morphology. It is shown that both hillock and ring arrays can be generated with well-controllable dimensions.

EXPERIMENTAL

Hexagonally close-packed monolayers of PS spheres were created exploiting self-organization effects along a receding triple-phase boundary of a droplet of colloidal suspension drying in a controlled fashion on a pre-cleaned Si(100) surface, as described in more detail in [2]. In order to be able to study the influence of the mask opening size on the height of surface features, commercially available aqueous suspensions of PS beads with diameters of 200, 600, and 1000 nm were used, which are expected to form triangular mask openings with concave sides and perpendicular bisector of the sides of 73, 220, and 365 nm length, respectively. The nanomasks were inspected by light and scanning electron microscopy (SEM) prior to irradiation with He ions in an Eaton NV3204 medium current ion implanter at room temperature. Ion doses of up to 0.1 to 2 x 10^{17} He/cm^2 were used. The ion energy was systematically varied between 8 and 48 keV in order to study the influence of the perpendicular and lateral ion range on the local surface swelling of Si wafers. According to static SRIM [7] Monte Carlo simulations assuming a PS density of 1.06 g/cm^3 the range of He ions varies between 124 and 534 nm, the longitudinal straggling between 41 and 82 nm and the lateral straggling between 36 and 94 nm. Corresponding He ion ranges and stragglings in Si are 82-395 nm, 51-120 nm, and 42-120 nm, respectively. NSL masks were removed using an adhesive tape prior to surface characterization with an AFM in non-contact mode, SEM and cross-sectional transmission electron microscopy (XTEM) in a Jeol 2100 F at 200 kV.

RESULTS AND DISCUSSION

Fig. 1 shows an SEM image of the boundary between the irradiated (left) and the un-irradiated part (right) of a NSL mask of 1000 nm PS beads. Some linear defects are visible already in the un-irradiated part of the mask, sometimes being related to a point defect in the mask (arrow in Fig. 1) and sometimes apparently being due to drying effects in the mask. Owing to beam induced charging the linear defects widen up considerably, as one can clearly see at boundaries of the irradiated mask area as in Fig. 1. Even though this beam charging induced effect is not desirable it allows us to study the behavior at large mask openings at the same time.

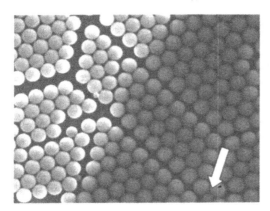

Figure 1. SEM secondary electron image of the boundary between the un-irradiated (right) and the irradiated part (left) of a NSL mask made of 1000 nm diameter PS beads. Implantation was done at an energy of E = 8 keV and a dose of D = 2 · 10^{17} He$^+$/cm^2. The arrow points to a point defect of the mask consisting of an overly small PS bead and most likely causing an extended linear defect running to the edge of the implanted area and expanding into a much wider defect on the irradiated side (left).

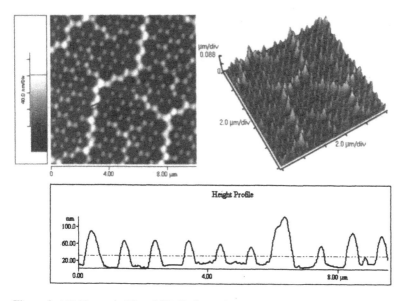

Figure 2. AFM image in 2D and 3D display and corresponding height profile taken on the irradiated side of the sample of Fig. 1 after removing the NSL mask.

The local swelling induced by He ion implantation through such a 1000 nm PS bead NSL mask at a dose of 2×10^{17} He/cm^2 is shown in the AFM images in Fig. 2. At each of the triangular mask openings a sharply pointed tip has developed. The height of these tips is relatively uniform; however, at the linear mask defects, much higher protrusions have developed, confirming a trend also observed in our previous studies [6] on NSL masks from 200 nm diameter PS beads.

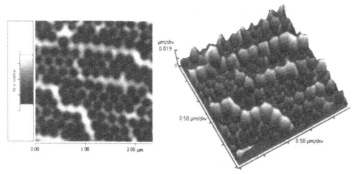

Figure 3. AFM image displayed in 2D and 3D showing the surface morphology of silicon after implantation of 8 keV He ions at a dose of $D = 1 \cdot 10^{17}$ He$^+$/cm^2 after removal of the 200 nm bead diameter NSL mask.

Arrays of trough-like nanostructures are obtained instead of arrays of individual hillocks if similar implantations are performed through NSL masks with smaller spheres. This is demonstrated in Fig. 3 showing the result of an implantation at the same energy as in Fig. 2 but with 200 nm spheres instead of 1000 nm diameter. Even though the dose is smaller by a factor of two in Fig. 3 compared to the sample in Fig. 2, the hillocks have started to merge into rings.

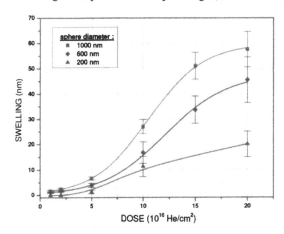

Figure 4. Average feature height of swollen hillocks as a function of dose of 8 keV He ions for three different NSL mask sizes, 200, 600 and 1000 nm. Error bars represent the standard deviations of AFM measured dot heights.

The dose dependence of the average feature heights measured by AFM in areas without mask defects is displayed in Fig. 4 for masks made out of 200, 600, and 1000 nm diameter PS beads and a fixed implantation energy of 8 keV. While our preliminary studies in [6] using 200 nm spheres have indicated a roughly linear dose dependence of the average feature heights, a more complicated behavior becomes obvious in particular for the NSL masks made out of larger spheres. The feature height generally follows an s-shaped dependence as a function of dose, with larger heights reached for larger mask openings. The s-shape of curves may be explained by the necessity to overcome an incubation dose before a measurable swelling sets in. In fact, for 200 nm sphere masks no measurable swelling is present at the lowest doses ($1-2 \cdot 10^{16}$ He$^+$/cm^2), while it is clearly detectable for the larger mask openings at the same doses. This and also the subsequent smaller height increase at higher doses in the case of smaller mask openings can be attributed to the fact that owing to the lateral straggling of He ions in Si (42 nm at 8 keV) the local dose reached in the centre of the mask opening is the smaller, the narrower the mask opening is. Another reason for the mask size dependence of the feature height is of course the larger stability of the surface against swelling if the fixed, un-swollen rims of the implanted patches are close by.

Surprisingly there is obviously also some saturation of the height increase with increasing dose (Fig. 4). If one assumes that the swelling volume is proportional to the local concentration of He atoms, then a reduced increase of feature heights per dose should be accompanied by a larger lateral expansion of the swollen volumes. In fact one can observe both by AFM and SEM that at the highest doses the basis of hillocks has turned from a concave to a convex shape with some of neighbouring hillocks starting to get interconnected. It should be noted that along linear mask defects, up to the highest doses used, the feature heights do not saturate but increase up to 140nm

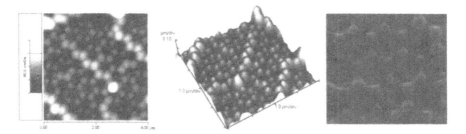

Figure 5. AFM image displayed in (a) 2D and (b) 3D and (c) SEM image of the surface of Si implanted through a 600 nm diameter NSL mask with 8 keV He ions at a dose of $D = 2 \cdot 10^{17}$ He^+/cm^2. The scanning area of the AFM images is 4 x 4 μm^2, while the SEM image is about 2 x 2 μm^2.

(not shown), indicating that the fixation of expanding volumes by the rims of the features has a marked effect on the obtainable feature heights. In the absence of such rims at large mask openings or especially along the broad linear mask defects the highest protrusions are therefore obtained.

The inner structure of a hillock obtained by implantation through a 600 nm mask is displayed in Fig. 6. It is obvious that hillocks are caused by the formation of ensembles of voids of different size, which are self-arranged in a hierarchical order with one large cavity in the centre and increasingly smaller ones in shells around the centre. The voids are embedded in amorphized Si, presumably allowing for a viscous flow.

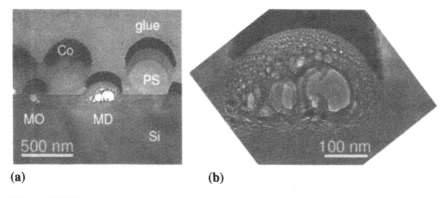

(a)   (b)

Figure 6. XTEM image of the Si surface locally implanted through a 600 nm diameter NSL mask with 8 keV He ions at a dose of $D = 2 \cdot 10^{17}$ He^+/cm^2. (a) Two hillocks obtained at a mask defect (MD) and a regular mask opening (MO) are shown. The structure is covered with Co, serving as a protection layer during XTEM specimen preparation, and glue. (b) Magnified view of the hillock obtained underneath the mask defect, clearly showing the hierarchical arrangement of voids of different size.

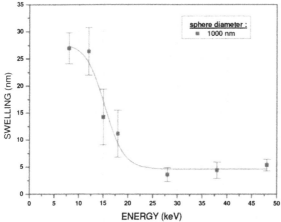

Figure 7. Average feature height of hillock arrays measured by AFM at the openings of 1000 nm diameter NSL masks as a function of He ion energy for a fixed dose of $D = 2 \cdot 10^{17}$ He$^+$/cm^2. Error bars represent the standard deviations of hillock heights.

Upon an increase of the ion energy, the thickness of the Si layer above the central cavity (the so called "Deckeldicke") must get larger. It may be expected that also residual crystalline material may be present near the surface at higher energies. Thus with increasing ion energy the average hillock height is expected to get smaller. In fact, this is observed (Fig. 7). A sharp decrease of the average hillock height is obtained between 10 and 20 keV and some levelling off at energies above 20 keV, even for a dose as high as $2 \cdot 10^{17}$ He$^+$/cm^2. Nevertheless, at the highest energies the curve does not level off at zero and some residual swelling can be measured. This probably indicates the presence of yet another swelling mechanism, possibly driven by out-diffusion of point defects. For a modelling of such behaviour, however, more details of the dose dependence of the structure of the Si capping layer are necessary, which requires further investigations.

CONCLUSIONS

Nanosphere lithography (NSL) in combination with high-dose He ion implantation is shown to be a promising new technique to form periodically nanostructured Si surfaces based on the well controllable localized swelling of He implanted near-surface regions.

REFERENCES

1. Ch. L. Haynes, R.P. Van Duyne, *J. Phys. Chem. B* 2001, *105*, 5599-5611.
2. J.K.N. Lindner, B. Gehl, B. Stritzker, Nucl. Instr. and Meth. B 242 (2006) 167.
3. J.K.N. Lindner, D. Kraus, B. Stritzker, Nucl. Instr. and Meth. B 257 (2007) 455.
4. G.F. Cerofolini et al., Mat. Sci. and Eng. 27 (2000) 1-52.
5. V. Raineri, S. Coffa, E. Szilágyi, J. Gyulai, E. Rimini, Phys. Rev. B 61 (2000) 937.
6. J.K.N. Lindner et al., Nucl. Instr. and Meth. B 267 (2009) 1394.
7. J.F. Ziegler, J.P. Biersack, U. Littmark, *The Stopping and Range of Ions in Matter*, Pergamon, New York, 1985; here the simulation code SRIM 2003.26 was used.

Mater. Res. Soc. Symp. Proc. Vol. 1181 © 2009 Materials Research Society   1181-DD07-01

Localized Gallium Doping and Cryogenic Deep Reactive Ion Etching in   Fabrication of Silicon Nanostructures

Nikolai Chekurov[1,2], Kestutis Grigoras[1,2], Antti Peltonen[1,3], Sami Franssila[1,2] and Ilkka Tittonen[1,2]

[1] Department of Micro and Nanosciences, Helsinki University of Technology,

PO Box 3500, FIN-02015 TKK, Finland

[2] Center for New Materials, Helsinki University of Technology, PO Box 3500, FI-02015 TKK, Finland

[3] TKK Micronova, Helsinki University of Technology, PO Box 3500, FI-02015 TKK, Finland

ABSTRACT

We present a novel fabrication method to create controlled 3-dimensional silicon nanostructures with the lateral dimensions that are less than 50 nm as a result of a rapid clean room compatible process. We also demonstrate periodic and nonperiodic lattices of nanopillars in predetermined positions with the minimum pitch of 100 nm. One of the uses of this process is to fabricate suspended silicon nanowhiskers.

INTRODUCTION

Instead of using focused ion beam (FIB) in milling the target by bombarding it with gallium ions, the doping of silicon is known to lead to selective masking of the surface in etching [1]. Especially the wet etch process [2, 3] is known to be sensitive for gallium ion doping. More recently, the dry processes have been studied in association with gallium implantation especially reactive ion etching with various chemical compositions and its derivatives such as deep reactive ion etching techniques (DRIE).

The main drawback of the reported methods is a poor selectivity between treated and untreated areas of the sample (dry etching) or crystallographic anisotropy restrictions (wet etching). In dry etching, the selectivity values in the range of 1-2.5 [4] have been demonstrated. In this work we describe a combination of local gallium implantation and cryogenic deep reactive ion etching which enables selectivity of least 2000:1 thus allowing fabrication of deep structures. By using the adjustable etching process one can achieve controlled underetching of the structures creating horizontally suspended nanowhiskers.

EXPERIMENT

The fabrication process consists of only two main steps (figure 1). First selected area of the sample is treated with Ga^+ ion beam (FEI Helios Nanolab 600) and then DRIE (Oxford Instruments Plasmalab System 100) is used to machine the features by removing the untreated silicon. In analyzing the gallium dose needed to protect silicon from etching in the cryogenic DRIE we found out that moderate amount of 10^{16} ions/cm^2 is enough to produce structures of several μm in height. The dose is several orders of magnitude lower than the one needed for traditional direct FIB milling or FIB – assisted etching (10^{16} ions/cm^2 instead of

> 10^{18} ions/cm^2). By altering the etching parameters during the DRIE step, we can choose between the vertical and anisotropic sidewall profile. This gives us a possibility to create freestanding silicon nanostructures such as isolated nanowires or complete networks of nano electro mechanical systems.

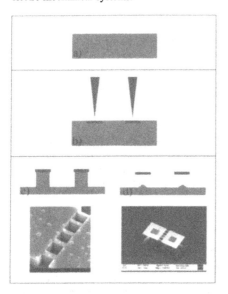

Figure 1. The method of gallium ion implantation for cryogenic deep reactive ion etching. a) plain silicon substrate b) patterning by local gallium doping c) anisotropic reactive ion etching d) isotropic reactive ion etching.

To determine the gallium doping dose required to protect silicon from etching a following experiment was undertaken: 45 150 x 150 μm^2 areas with the dose varying from $2.5 \cdot 10^{14}$ to $3.3 \cdot 10^{16}$ ions/cm^2 were patterned. The patterns were etched in DRIE for 1, 2 and 4 minutes using 800 W ICP and 3 W CCP power, 40 sccm SF$_6$ flow and 6 sccm O$_2$ flow in temperature of -120°C. The heights of the resulting structures were measured after every etching and results are shown in figure 2, where we can see that the doses above 10^{16} ions/cm^2 always produce the full height structure, meaning that the mask is durable enough to withstand the whole etching process. It is also evident that by using relatively small doses from 10^{15} to $7 \cdot 10^{15}$ ions/cm^2 and short etching times we can obtain structures of a varying height during a single etch step. It is important not to use much larger than the required dose even for full masking, as extra ions widen the implantation regions and decrease the accuracy in patterning.

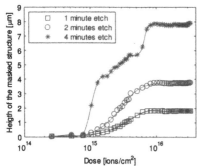

Figure 2. Influence of the amount of the gallium dose as a masking layer, the full height structures with doping dose $> 10^{16}$ ions/cm^2 indicate that the mask did not fail during etching. The structures with the gallium ion concentration $< 10^{16}$ ions/cm2 experience a mask failure which is more severe for longer etching times.

To determine the resolution of the process, we created arrays of nanopillars with masks of different shapes. For this experiment the smallest available ion current (1.5 pA) and the shortest dwell time (100 ns) available were utilized. The treatment was repeated for 40 cycles resulting in $4 \cdot 10^{16}$ ions/cm^2. Figure 3 shows such pillars with square masking as well as parts of electronic masks and resulting structures from the same viewing angle and in scale. The measured widening of the structures was from 17 nm to 25 nm. A minimum linewidth was measured by creating a line pattern with varying thickness and spaces between the lines. As a result, 43 nm wide trenches and 45 nm wide lines were obtained.

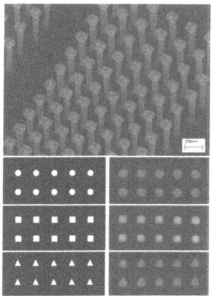

Figure 3. Nanopillar resolution test. Arrays of nanopillars (top) were fabricated, pictured directly from above (right) and compared to the starting mask (left). The widening of the structures measured to be in the range of 17-25 nm.

3D suspended structures can generally be fabricated with the same resolution as the non-suspended ones. Additionally, by using a relatively high etching temperature (-80 °C instead of -120°C), shrinkage of gallium doped layer occurs, making it possible to create nanobridges down to 20 nm wide (Figure 4).

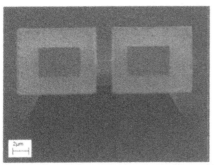

Figure 4. Smallest nanobridge obtained by etching at elevated (-80°C) temperature. The length of the bridge is 2 μm, width < 20 nm and thickness < 30 nm.

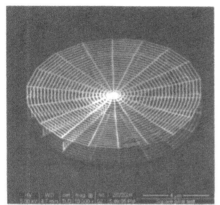

Figure 5. A floating web formation demonstrating toughness and extremely low stress of the resulting suspended nanowhiskers. The structure (10 μm in diameter) consists of crossing nanobridges from 50 nm to 100 nm wide. There is only little stress-related buckling visible.

DISCUSSION

Assuming that the penetration depth of the 30 keV gallium ions in silicon is around 30 nm, the selectivity between treated and untreated silicon is well over 2000 as structures up to 80 um high were obtained in the mask stress experiments. The lateral widening of the structures was measured to be between 17 nm and 25 nm and the overall process resolution is 10 pairs of lines / μm. An achievable height-to width aspect ratio is measured to be more than 15:1, which means that e.g. pillars 600 nm high and 40 nm in diameter are possible. The repeatability of the process within one run is excellent, as several thousands of identical structures can be produced (Figure 6). The processing time is limited by the speed of the FIB – writing, because the speed of the DRIE – step is more than sufficient; 2 μm/min, and thus typical etch times are well below one minute. The FIB step takes, depending on the resolution and treated area, from microseconds to several minutes. By using different FIB – currents and view fields it is possible to create structures at an area of 10 x 10 μm^2 even with 150 μm times 150 μm bonding pads in under 15 minutes, so several design/fabrication/measurement cycles can be accomplished in one working day. For the freestanding structures, our current achievement is a nanowire with dimensions of 2 μm x 20 nm x 30 nm.

Figure 6. An array of more than 1000 identical nanopillars 300 nm in diameter and 4 um high.

CONCLUSIONS

The fabrication method producing silicon structures with a line width under 50 nm and with an aspect ratio of more than 15:1 has been developed. This quick, all-dry process can be utilized for several purposes in nanotechnology research and prototyping phases. Possible research areas, which would greatly benefit from possibility to quickly realize nanometer-sized structures are plasmonics and metamaterial or quantum phenomena research.

ACKNOWLEDGMENTS

Nikolai Chekurov acknowledges Magnus Ehrnrooths foundation for financial support.

REFERENCES

[1] A. J. Steckl, H. C. Mogul, S. Mogren, Applied Physics Letters, **60**, 1883 (1992)

[2] J. Brugger, G. Beljakovic, M. Despont, N. F. De Rooij, P. Vettiger, Microelectronic engineering, **35**, 401 (1997)

[3] B. Schmidt, S. Oswald, L. Bischoff, J. of the Electrochemical Society, **152**, G875 (2005)

[4] H. X. Qian, W. Zhou, J. Miao, L. E. N. Lim, X. R. Zeng, Journal of Micromechanics and Microengineering, **18**, 35003 (2008)

Mater. Res. Soc. Symp. Proc. Vol. 1181 © 2009 Materials Research Society   1181-DD07-03

Ion Implanted SiO₂ Substrates for Nucleating Silicon Oxide Nanowire Growth

Jason L. Johnson, Yongho Choi, and Ant Ural

Department of Electrical and Computer Engineering, University of Florida, Gainesville, Florida 32611, USA

ABSTRACT

We experimentally demonstrate a simple and efficient approach for silicon oxide nanowire growth by implanting Fe^+ ions into thermally grown SiO_2 layers on Si wafers and subsequently annealing in argon and hydrogen to nucleate silicon oxide nanowires. We study the effect of implantation dose and energy, growth temperature, and H_2 gas flow on the SiO_x nanowire growth. We find that sufficiently high implant dose, high growth temperature, and the presence of H_2 gas flow are crucial parameters for silicon oxide nanowire growth. We also demonstrate the patterned growth of silicon oxide nanowires in localized areas by lithographic patterning and etching of the implanted SiO_2 substrates before growth. This works opens up the possibility of growing silicon oxide nanowires directly from solid substrates, controlling the location of nanowires at the submicron scale, and integrating them into nonplanar three-dimensional nanoscale device structures.

INTRODUCTION

One-dimensional (1D) nanostructures, such as nanotubes and nanowires, have attracted significant research attention in recent years due to their unique structural and electronic properties. Controlled growth and synthesis of such 1D nanostructures remain an active research area. SiO_2 is a material which is of great technological importance in silicon VLSI technology. Nanowires of silicon oxide have a great potential in applications such as low dimensional waveguides, scanning near-field optical microscopy, blue light emitters, nanoscale optical devices and sensors, sacrificial templates, and biosensors [1-3]. Several methods have been used to grow silicon oxide nanowires, such as laser ablation [2], thermal evaporation [4], and chemical vapor deposition (CVD) [5]. In most of these cases, a growth model based on the vapor-liquid-solid (VLS) growth mechanism [6] has been used to explain the observed results.

An essential component of the VLS growth process is the nanoscale catalyst particles required to nucleate the growth of nanowires. For example, several recent studies have demonstrated the growth of silicon oxide nanowires from a variety of different catalyst nanoparticles, including sputtered or evaporated metal thin films [1,7] and molten Ga [5]. In some of these studies, Si was supplied as a powder [2, 4] or in gaseous phase as silane (SiH_4) [5]. In another study, the catalyst material was deposited directly on the Si substrate [1].

Nanoscale catalyst particles formed from deposited thin films and powders are typically difficult to pattern into very small dimensions or into nonplanar three-dimensional (3D) device structures, such as the sidewalls of high aspect ratio trenches [8]. An alternative catalyst deposition technique, which has not been studied as much [3, 8-11], is to use ion implantation and subsequent annealing to create catalyst nanoparticles.

A few previous experiments have shown that ion implantation and subsequent annealing can create catalyst nanoparticles for nucleating the growth of multi-walled carbon nanotubes (MWNTs) [8-10]. More recently, the growth of silicon nanowires by CVD on gold implanted silicon substrates was demonstrated [12, 13]. In another recent work, palladium ion implantation into bare Si wafers was used to grow silica nanowires [3]. Furthermore, we have recently shown that single-walled carbon nanotubes (SWNTs) [11], GaN nanowires, and Ga_2O_3 nanowires and nanoribbons [14] can be produced by the process of iron (Fe) ion implantation into thermally grown SiO_2 layers, subsequent annealing, and CVD growth.

In this paper, we experimentally demonstrate a simple and efficient approach for silicon oxide nanowire growth by implanting Fe^+ ions into thermally grown SiO_2 layers on Si wafers to nucleate silicon oxide nanowires during subsequent annealing in argon and hydrogen. In contrast to most previous work, both reactants (Si and O) come from the SiO_2 substrate, which acts as a solid source. Our results simultaneously show that ion implantation can be used as a versatile nucleation method, iron (a commonly used catalyst for carbon nanotube growth) is an efficient catalyst, and that SiO_2 layers can be used as a solid source for silicon oxide nanowire growth [15].

EXPERIMENTAL METHODS

In our experiment, 500 nm thick SiO_2 layers were first thermally grown on silicon (100) substrates by wet oxidation. Fe^+ ions were implanted into these layers, as depicted in Fig. 1(a), at an energy of 60 keV with three different doses (10^{14}, 10^{15}, and 10^{16} cm^{-2}) and at a dose of 10^{15} cm^{-2} with three different energies (25, 60, and 130 keV). The projected range R_p of the 25, 60, and 130 keV energy implants in SiO_2 are 24, 50, and 103 nm, respectively, based on SRIM [16] calculations. The implant energies were chosen such that most of the as-implanted Fe atoms are confined close to the substrate surface. The depth profile of the 60 keV, 10^{15} cm^{-2} implant calculated by SRIM simulations is shown in Fig. 1(b), where the peak concentration C_p is ~1.5×10^{18} cm^{-3}. The implant doses were chosen based on our previous results for carbon nanotube, and GaN and Ga_2O_3 nanowire growth [11, 14].

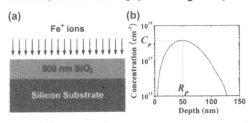

Figure 1. (a) The schematic of the SiO_2/Si substrate and Fe ion implantation used for the catalyst nanoparticle formation for silicon oxide growth. (b) The depth profile of Fe atoms ion-implanted into thermally grown SiO_2 at an energy of 60 keV and a dose of 10^{15} cm^{-2}, calculated by SRIM simulations, giving a projected range of $R_p = 50$ nm and a peak concentration of $C_p \sim 1.5 \times 10^{18}$ cm^{-3}, as labeled in the figure.

For unpatterned growth of silicon oxide nanowires, the as-implanted samples were placed in a one inch quartz tube furnace. The quartz tube was then purged at room temperature with 350 sccm flow rate of Ar and 200 sccm flow rate of H_2 for 15 mins. After purging the tube, the temperature was increased to 1100°C and the samples were annealed for 30 mins under the same gas flow rates (350 sccm Ar and 200 sccm H_2) to grow the silicon oxide nanowires.

For patterned growth, the 60 keV, 10^{15} cm^{-2} as-implanted samples were patterned by standard photolithography and etched using BOE, such that no SiO_2 remained in the etched areas. Since most of the implanted Fe atoms are contained in the top ~150 nm of the SiO_2 layer for this case [see Fig. 1(b)], etching the SiO_2 in selective areas removes all catalyst in those areas. After the patterning step, these samples were grown using the same conditions given above for the unpatterned samples. The as-grown samples were characterized by scanning electron microscopy (SEM), high resolution transmission electron microscopy (HRTEM) with selected area electron diffraction (SAED) and energy dispersive X-ray spectroscopy (EDS) capability, and atomic force microscopy (AFM).

RESULTS & DISCUSSION

The SEM images of Figs. 2(a), (b), and (c) show the unpatterned growth results from 60 keV implants with 10^{14}, 10^{15}, and 10^{16} cm^{-2} dose, respectively. As seen in Fig. 2(a), no nanowires were grown on the low dose sample, whereas the 10^{15} and 10^{16} cm^{-2} dose samples showed dense growth of long nanowires. Using a simple model, the flux of Fe atoms diffusing to the oxide surface is proportional to the dose of the implant; as a result, a low dose results in insufficient size and density of catalyst nanoparticles and no nanowire growth.

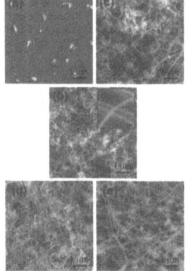

Figure 2. Scanning electron microscopy (SEM) images of silicon oxide nanowires grown from Fe catalyst ion implanted into thermally grown SiO_2 layers. The implant energies and doses are (a) 60 keV, 10^{14} cm^{-2}, (b) 60 keV, 10^{15} cm^{-2}, (c) 60 keV, 10^{16} cm^{-2}, (d) 25 keV, 10^{15} cm^{-2}, and (e) 130 keV, 10^{15} cm^{-2}, respectively. The inset of part (b) shows the close-up of nanowires with diameters of ~40 nm. The scale bar in the inset is 100 nm.

The SEM images of Figs. 2(d), (b), and (e) show the unpatterned growth results from 10^{15} cm^{-2} dose implants with 25, 60, and 130 keV energy, respectively. Nanowire growth is observed for all energies; however, the nanowires on the 130 keV sample look more straight, most likely due to the lower nanowire density caused by the deeper location of catalyst in the SiO_2 layer (i.e. larger R_p) for that case. Furthermore, based on the analysis of SEM and TEM images, the diameters of the as-grown silicon oxide nanowires were found to be in the range between 10 and 50 nm [see inset of Fig. 2(b)]. The as-grown nanowires from the 60 keV, 10^{15} cm^{-2} implant were also characterized by HRTEM. Figures 3(a) and (b) show low resolution and high resolution TEM images, respectively, of the silicon oxide nanowires. Quantitative EDS analysis revealed an atomic ratio of Si to O of 1:2.6. We performed growth on a control SiO_2 substrate with no Fe ion implantation. The growth temperature was 1100°C and all the other growth parameters were the same as before. No nanowire growth was observed in the absence of Fe ion implantation. These results provide clear evidence that the presence of Fe is essential for the growth of silicon oxide nanowires.

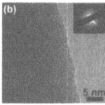

Figure 3. (a) Low resolution and (b) high resolution TEM images of silicon oxide nanowires. The inset of (b) shows the selected area electron diffraction of an individual nanowire. The high resolution TEM image and the SAED pattern confirm that the SiO_x nanowires are amorphous.

In order to gain a further understanding of the effect of growth parameters on the silicon oxide nanowire growth, the effect of growth temperature and H_2 gas flow was also investigated for the 60 keV, 10^{15} cm^{-2} implant sample. First, we report on the effect of temperature. The unpatterned growth was repeated by lowering the temperature to 1000°C, keeping all the other parameters constant, and we found that no nanowire growth was observed at this temperature. Next, we investigate the role of H_2 gas flow in the nanowire growth process. For this case, the growth temperature was kept at 1100°C, but no H_2 gas was supplied during the growth process; we found that no nanowire growth was observed in the absence of H_2 flow. Hydrogen is known to enhance the diffusion of impurities in SiO_2. An increase in the catalyst nanoparticle density was observed when H_2 is flown during the growth, which can be explained by the enhancement of Fe diffusion in SiO_2 in H_2 ambient. Furthermore, SiO_2 can be reduced by hydrogen, which supplies more reactants of Si and O to the nanowire. As a result, the presence of H_2 gas flow is also a crucial factor for silicon oxide nanowire growth.

Finally, we present the results of patterned growth from the 60 keV, 10^{15} cm^{-2} implant sample. After implantation, the SiO_2 layer was patterned by photolithography and etched by BOE, as explained previously. The sample was grown at 1100°C using the same standard recipe as in the unpatterned case [See Fig. 2(b)]. The SEM images of Figs. 4(a) and (b) show the growth results. It is clear from Figs. 4 that the silicon oxide nanowires have grown selectively only from the unetched areas of the patterned sample. These results demonstrate that ion implantation-based nucleation methods can be used for patterned growth of silicon oxide nanowires only in localized areas on a silicon substrate.

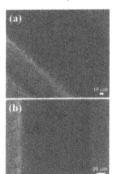

Figure 4. SEM images of patterned growth from the 60 keV, 10^{15} cm^{-2} implant sample. The silicon oxide nanowires have grown selectively only from the unetched areas of the patterned SiO_2 sample.

The vapor-liquid-solid (VLS) growth mechanism [6] has been commonly used to explain the growth of nanowires from catalyst nanoparticles. In our silicon oxide nanowire growth process, silicon and oxygen are provided directly from the solid SiO_2 substrate. The physical model for the growth of silicon oxide nanowires can be proposed as follows: First, the as-implanted Fe [Fig. 5(a)] out-diffuses during the annealing to form catalyst nanoparticles on the SiO_2 surface [Fig. 5(b)]. The catalyst nanoparticles from which nanowire growth has not yet nucleated could also get larger by a surface diffusion and Ostwald ripening process. Droplets containing Fe form from these catalyst nanoparticles. When these droplets become supersaturated in silicon and oxygen, silicon oxide nanowires nucleate and precipitate out of these droplets [Fig. 5(c) and (d)].

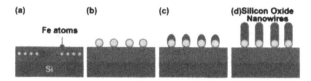

Figure 5. The schematic diagram of the proposed growth model of silicon oxide nanowires: (a) First, Fe atoms are ion-implanted into the SiO_2 substrate. (b) At high temperature, Fe out-diffuses to form catalyst nanoparticles on the SiO_2 surface. (c), (d) Droplets containing Fe form from these catalyst nanoparticles. When these droplets become supersaturated in silicon and oxygen, silicon oxide nanowires nucleate and precipitate out of these droplets.

CONCLUSIONS

In conclusion, we have experimentally demonstrated the patterned growth of silicon oxide nanowires from Fe^+ ions implanted into thermally grown SiO_2 layers. We have also studied the effect of implantation dose and energy, growth temperature, and the presence of H_2 gas flow on the silicon oxide nanowire growth. TEM, SAED, and EDS characterization revealed that the as-grown nanowires are amorphous silicon oxide with a Si:O ratio of 1:2.6. Furthermore, sufficiently high implant dose, high growth temperature, and the presence of H_2 gas flow were found to be crucial parameters for silicon oxide nanowire growth. We also demonstrated the patterned growth of silicon oxide nanowires in localized areas by lithographic patterning and etching of the implanted SiO_2 substrates before growth. We proposed a simple physical model to explain the growth results. This method of nucleating nanowire growth is not limited to silicon oxide nanowires; it could also be generally applied to the growth of other types of nanowires for potential nanoscale device applications.

ACKNOWLEDGMENTS

The authors would like to thank Kerry Siebein for her assistance with TEM characterization. Fe ion implantation was performed at Core Systems, Sunnyvale, CA, USA. The AFM, SEM, and TEM characterization was done at the Major Analytical Instrumentation Center (MAIC) at the University of Florida. This work was funded in part by the University of Florida Research Opportunity Fund.

REFERENCES

1. Wang C Y, Chan L H, Xiao D Q, Lin T C and Shih H C 2006 J. Vac. Sci. Technol. B 24 613
2. Yu D P, Hang Q L, Ding Y, Zhang H Z, Bai Z G, Wang J J, Zou Y H, Qian W, Xiong G C and Feng S Q 1998 Appl. Phys. Lett. 73 3076
3. Sood D K, Sekhar P K and Bhansali S 2006 Appl. Phys. Lett. 88 143110
4. Liang C H, Zhang L D, Meng G W, Wang Y W and Chu Z Q 2000 J. Non-Cryst. Solids 277 63
5. Pan Z W, Dai S, Beach D B and Lowndes D H 2003 Nano Lett. 3 1279
6. Wagner R S and Ellis W C 1964 Appl. Phys. Lett. 4 89
7. Park H K, Yang B, Kim S W, Kim G H, Youn D H, Kim S H and Maeng S L 2007 Physica E 37 158
8. Adhikari A R, Huang M B, Wu D, Dovidenko K, Wei B Q, Vajtai R and Ajayan P M 2005 Appl. Phys. Lett. 86 053104
9. Mao J M, Sun L F, Qian L X, Pan Z W, Chang B H, Zhou W Y, Wang G and Xie S S 1998 Appl. Phys. Lett. 72 3297
10. Ding X Z, Huang L, Zeng X T, Lau S P, Tay B K, Cheung W Y and Wong S P 2004 Carbon 42 3030
11. Choi Y H, Sippel-Oakley J and Ural A 2006 Appl. Phys. Lett. 89 153130
12. Stelzner T, Andra G, Wendler E, Wesch W, Scholz R, Gosele U and Christiansen S 2006 Nanotechnology 17 2895
13. Christiansen S, Schneider R, Scholz R, Gosele U, Stelzner T, Andra G, Wendler E and Wesch W 2006 J. Appl. Phys. 100 084323
14. Johnson J L, Choi Y and Ural A 2008 J. Vac. Sci. Technol. B 26 1841
15. Choi Y, Johnson J L, and Ural A 2009 Nanotechnology 20 135307
16. Ziegler J F and Biersack J P computer code SRIM (www.srim.org).

Magnetic, Optical and
Semiconductor Applications

Mater. Res. Soc. Symp. Proc. Vol. 1181 © 2009 Materials Research Society 1181-DD02-03

Tunneling and anisotropic-tunneling magnetoresistance in iron nanoconstrictions fabricated by focused-ion-beam

Amalio Fernández-Pacheco[1,2,3], José M. De Teresa[2,3], R. Córdoba[1,3] and Ricardo Ibarra[1,2,3]
[1] Instituto de Nanociencia de Aragón, Universidad de Zaragoza, Zaragoza, 50009, Spain
[2] Instituto de Ciencia de Materiales de Aragón, Universidad de Zaragoza-CSIC, Facultad de Ciencias, Zaragoza, 50009, Spain
[3] Departamento de Física de la Materia Condensada, Universidad de Zaragoza, Facultad de Ciencias, Zaragoza, 50009, Spain

ABSTRACT

We report the magnetoresistance (MR) measurements in a nanoconstriction fabricated by focused-ion-beam (FIB) in the tunneling regime of conductance. The resistance of the contact was controlled during the fabrication process, being stable in the metallic regime, near the conductance quantum, and under high vacuum conditions. The metallic contact was deteriorated when exposed to atmosphere, resulting in a conduction mechanism by tunneling. The TMR was found to be of 3% at 24 K. The anisotropic tunneling magnetoresistance (TAMR) was around 2% for low temperatures, with a field angle dependence more abrupt than in bulk Fe. This preliminary result is promising for the application of this technique to fabricate stable ferromagnetic constrictions near the atomic regime of conductance, where high MR values are expected.

INTRODUCTION

The mechanism of electronic transport in constrained geometries on the nanometer scale changes dramatically in comparison with bulk, once the dimensions are reduced to less than the mean free path of electrons. In this regime, electronic transport is not diffusive anymore, but ballistic, and the conductance can become quantized ($G=nG_0$, where $G_0=2e^2/h$ is the conductance quantum) [1]. In the case of magnetic materials, high-impact results were reported in the past in atomic-size contacts, with extremely large values for MR, phenomenon coined as "ballistic MR" (BMR) [2,3]. This effect was explained by the pinning of a domain wall in the constriction, restricting the transmission of electrons. However, subsequent experiments revealed that mechanical artifacts were playing a major role for these large ratios, resulting in a huge controversy [4]. Another different finding in these structures was a large anisotropy in the MR of the contacts, whose magnitude and angular dependence were found to be very distinct from bulk materials, the so called "ballistic anisotropic MR" (BAMR) [5,6,7]. This phenomenon seems to be better established than the BMR, although critical voices claim that atomic reconfigurations in the contact, rather than an intrinsic electronic effect, could explain this behavior [8]. For a recent review in the topic see reference 9. It is therefore crucial, for the study of these effects, the fabrication of stable nanocontacts, where mechanical artifacts are avoided. Most of the work in this field has been done by techniques such as scanning tunneling microscope, mechanical break junction (MBJ) and electrochemical junctions [1,9]. New nanolithography techniques such as FIB [10,11] and electron beam lithography [12,13] have been recently used to fabricate nanoconstrictions, opening an interesting new route in this field, since all the structure, including the contact, is attached to the substrate. Thus, mechanical artifacts are minimized, and the developing of devices based on these effects would be, in principle, feasible. However, the fabrication of constrictions by these techniques in the sub-100 nm range is extremely difficult, and the creation of atomic-size constrictions is especially challenging. We recently demonstrated a new way to fabricate stable atomic-size constrictions, by the control of the resistance while the FIB etching is performed [14,15]. In this work we show the first magnetoresistance results in one iron nanoconstriction obtained by this method.

EXPERIMENT

The constriction was fabricated at room temperature in a commercial dual-beam equipment (Nova 200 NanoLab from FEI). A previous optical lithography step was carried out before the etching process, patterning large 10 nm-thick Fe pads on top of SiO_2. The Ga-FIB etching was done in a region of around 2×6 μm^2 (red square in Fig. 1a), with a beam acceleration voltage of 5 kV and a beam current of 10 pA. Two electrical microprobes by Kleindiek were contacted on the pads (see Fig.1a). These conductive microprobes are connected via a feedthrough to a 6220 DC current source- 2182A nanovoltmeter combined Keithley system, located out of the chamber, measuring the resistance of iron *in situ* while the FIB etching is being performed. The etching process lasts a few minutes, roughly corresponding to a total ion dose of 10^{17} ions/cm^2. In figure 1b) the microstructure of the constriction after the milling is shown and in the inset the conduction channel before the etching process. For experimental details see reference 14.

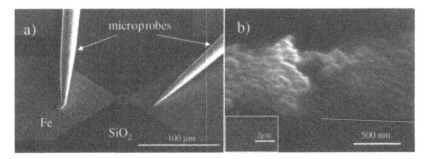

Figure 1. 52° tilted-view SEM images of the fabricated constriction. **a)** Experimental configuration, with the two microprobes contacted to the Fe pads for the *in situ* control of the resistance while the process is taking place. The red square indicates the etched zone. **b)** Microstructure of the iron after etching. A constriction is formed as a consequence of the ionic bombardment. The inset shows the metallic channel before the etching.

RESULTS

In figure 2 we show the evolution of the resistance with the process time. Since the initial resistance of the iron electrode is $R_i \sim 3$ kΩ, the resistance corresponding to the etched part (R_c) is the measured value minus R_i ($R=R_i + R_c$). After 2 minutes of etching, R starts to increase abruptly, and the FIB column is stopped when $R_c \approx 8$ kΩ. The R is measured during several minutes, finding that the constriction, in the metallic regime, is stable under the high vacuum conditions of the chamber (P~10^{-6} mbar). To avoid the possible deterioration when exposed to ambient conditions, a ~10 nm-thick layer of Pt-C was deposited by electrons (FEBID) on top of the etched zone, using $(CH_3)_3Pt(CpCH_3)$ as gas precursor. The high resistance of this material, because of the high amount of carbon, guarantees a resistance in parallel to the constriction of the order of tens of MΩ [16], which is confirmed by the negligible change of R while the deposit is done.

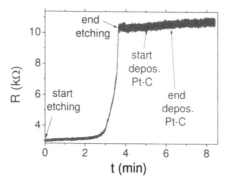

Figure 2. Resistance of the iron electrode as a function of the process time. The FIB etching is stopped when the resistance reaches 11 kΩ (R_c~8 kΩ). The constriction is covered by a thin layer of Pt-C, deposited by FEBID. The fabricated nanostructure is stable inside the vacuum chamber.

Once the constriction had been fabricated, the sample was put in contact with the atmosphere during some minutes, and transferred to a closed cycle refrigerator that can reach a minimum temperature of 24 K, combined with an electromagnet. The measurements of the resistance at room temperature showed an increase of the resistance by a factor of 10 (R≈100 kΩ), evidencing the departure from metallic conduction in the nanoconstriction. This value is significantly above the resistance corresponding to the quantum of conductance ($R_0=1/G_0=12.9$ kΩ). The slight non-linearity of the current-versus-voltage curves suggests that the constriction is not anymore in the metallic regime, but in the tunneling one (see Fig. 3a). The high reactivity of the nanostructure due to its large surface-to-volume ratio seems to be the most reasonable explanation of this effect. The resistance increased up to 120 kΩ ($G>9G_0$) at T=24 K.

We measured the MR for low temperatures with the magnetic field (H) parallel to the current path. In figure 3b the evolution with H and T is shown. MR ratios of the order of 3.2% are obtained. This value is a factor 30 times higher than in Fe non-etched samples, used as reference, where we observed a MR = -0.11 % in this configuration (note the different sign). As it is schematically explained by the red arrows in the graph, the evolution of R is understood by a change from a parallel (P) to an anti-parallel (AP) configuration of the ferromagnetic electrodes, separated by an insulator. More in detail, starting from saturation, a first continuous increase in R is observed, of the order of 0.8 %. This seems to be caused by a progressive rotation of the magnetization (M) in one of the electrodes. At around H=700 Oe, an abrupt jump of the resistance occurs (MR =2.5%), since the magnetization of one electrode switches its direction, and aligns almost AP to the M of the other electrode (intermediate state: IS). The MR becomes maximum at H=1 kOe, when the magnetization in both electrodes is AP. At H≈2 kOe, the M in the hard electrode also rotates, resulting again in a low-R state (P). As T increases, the IS, previously explained, disappears. IS is probably caused by the pinning of the M by some defects present in the soft electrode. Thus, the increase of the thermal energy favors the depinning of M. When the temperature was increased above 35 K, the nanoconstriction became degraded, likely due to the pass of the electrical current.

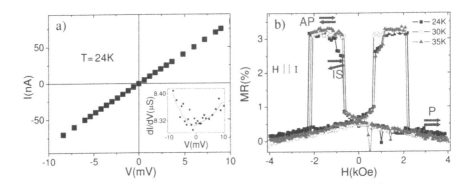

Figure 3. (a) Current-versus-voltage measurement at 24K. In the inset, the differential conductance of the curve deviates from the linear behavior, suggesting a non-metallic conduction. The line is a guide to the eye. **(b)** MR as a function of the magnetic field, applied parallel to the current direction. The MR is positive, with a value of 3.2% at 24 K, and can be understood by the decoupling of the magnetic electrodes separated by a tunneling barrier.

In figure 4a) the dependence of the MR at 24 K as a function of the voltage applied is shown, for the P and AP configurations. The diminishment of the resistance with the voltage is a typical feature of tunnel junctions, attributed to several factors such as the increase of the conductance with bias, excitation of magnons, or energy dependence of spin polarization due to band structure effects [17]. As the magnetostrictive state of parallel and antiparallel electrodes is the same, it seems that magnetostriction is not the cause of the observed TMR effect.

We have also studied the dependence of the MR with the field angle at saturation, by rotating the sample at the maximum field attained, H=11 kOe. In figure 1b) the evolution of MR is shown as a function of θ, the angle formed between H and the substrate plane. An anisotropic magnetoresistance effect (AMR) is present in the tunneling regime (TAMR). This effect is around 2% at 24 K, higher and of different sign from the AMR occurring in the bulk material (~ -0.3%). The TAMR has been previously observed in iron nanocontacts fabricated by MBJ [14], and implies that the evanescent wave functions maintain a strong atomic orbital character. The angle dependence is found to be more abrupt than the normal one for the AMR, proportional to $\cos^2\theta$. This behavior is typical for BAMR [5,14], and can be understood by considering the details of orbitals overlap [6,7].

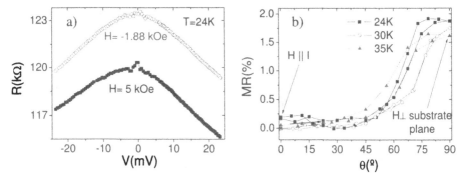

Figure 4. a) Bias dependence of the magnetoresistance at T= 24K for AP (H= -1.88 kOe) and P (H= 5 kOe) configurations. **b)** Angle dependence of the MR for low temperatures. A fixed field, H=11 kOe, is applied.

The experiments as a function of angle are also helpful to discard that magnetostriction effects are responsible of the observed MR. The magnetostriction in iron is given by

$$\frac{\Delta l}{l} = \frac{3}{2}\lambda_S (\cos^2\theta - \frac{1}{3}) \qquad (1)$$

with $\Delta l/l = -7 \times 10^{-6}$ for $\theta = 0°$, and $\Delta l/l = 3.5 \times 10^{-6}$ for $\theta = 90°$.

Thus, the increase of the MR as a function of θ has a contrary sign that if it was caused by magnetostriction, where the minimum resistance would be expected when H is perpendicular to the substrate plane.

CONCLUSIONS

In this contribution, we have shown the possibility to fabricate magnetic nanoconstrictions near the ballistic regime of conductance by the method we developed previously, using a FIB [14]. The constrictions are stable at room temperature and under high vacuum conditions. However, we have shown in this work the difficulty to have stable nanocontacts when exposed to ambient conditions.

An iron constriction in the tunneling regime, presents TMR ratios at low temperature 30 times higher than non-etched samples. We also observe a TAMR effect at low temperatures, of the order of 2%, and with an angle dependence different from the $\cos^2\theta$ expected for bulk AMR. From these experiments we also assure that magnetostriction is not playing any role in the measurements.

This result evidences that under low voltages etching, and with a moderated ion dose, the FIB procedure does not destroy the magnetic properties of the devices, although more research is required to investigate at what extent they are affected. The high stability expected for these constrictions in comparison with other suspended atomic-size structures, makes this first result promising for future research. A systematic study of the MR is currently under progress to improve the stabilization of constrictions in the metallic range of resistances, and investigate if high BMR and BAMR values can be attained at room temperature, which would have high impact in the field.

ACKNOWLEDGEMENTS

Financial support by the Spanish Monistry of Science (through project MAT2008-06567-C02, including FEDER funding), and the Aragon Regional Government are acknowledged.

REFERENCES

1. N. Agrait, A. L. Yeyati, and J. M. Van Ruitenbeek, *Phys. Rep.* **377**, 81 (2003).
2. N. García, M. Muñoz, and Y. W. Zhao, *Phys. Rev. Lett.* **82**, 2923 (1999).
3. J. J. Verlslujis, M. A. Bari, and J. M. D. Coey, Phys. Rev. Lett. **87**, 026601 (2001).
4. W. F. Egelhoff, Jr., L. Gan, H. Ettedgui, Y. Kadmon, C. J. Powell, P. J. Chen, A. J. Shapiro, R. D. McMichael, J. J. Mallett, T. P. Moffat, and M. D. Stiles, E. B. Svedberg, J. Appl. Phys **95**, 7554 (2004).
5. A. Sokolov, C. Zhang, E. Y. Tsymbal, J. Redepenning, and B. Doudin, *Nat. Nanotechnology* **2**, 171 (2007).
6. J. Velev, R. F. Sabirianov, S. S. Jaswal, and E. Y. Tsymbal, Phys. Rev. Lett. **94**,127203 (2005).
7. D. Jacob, J. Fernández-Rossier, and J. J. Palacios, Phys. Rev. B **77**, 165412 (2008).
8. S. –F. Shi, and D. C. Ralph, Nat. Nanotech. **2**, 522 (2007).
9. B. Doudin, and M. Viret, *J. Phys.: Cond. Matter.* **20**, 083201 (2008).
10. O. Céspedes, S. M. Watts, and J. M. D. Coey, *Appl. Phys. Lett.* **87**, 083102 (2005).
11. S. Khizroev, Y. Hijaki, R. Chomko, S. Mukheriee, R. Chantrell, X. Wu, R. Carley, D. Litvinov, *Appl. Phys. Lett.* **87**, 083102 (2005).
12. P. Krzysteczko, and G. Dumpich, *Phys. Rev. B* **77**, 144422 (2008).
13. T. Huang, K. Perzlmaier, and C. H. Back, *Phys. Rev. B* **79**, 024414 (2009).
14. A. Fernández-Pacheco, J. M. De Teresa, R. Córdoba, and M. R. Ibarra, *Nanotechnology* **19**, 415302 (2008).
15. J. V. Oboňa, J. M. De Teresa, R. Córdoba, A. Fernández-Pacheco, and M. R. Ibarra, *Microel. Eng.* **86**, 639 (2009).
16. J. M. De Teresa, R.Córdoba, A. Fernández-Pacheco, O. Montero, P. Strichovanec, and M.R. Ibarra, *J. Nanomat.* **2009**, 936863 (2009).
17. J. S. Moodera, and G. Mathon, *J. Magn. Mag. Mat* **200**, 248-273 (1999).
18. M. Viret, M. Gauberac, F. Ott, C. Fermon, C. Barreteau, G. Autes, and R. Guirado López, *Eur. Phys. J. B* **51**, 1 (2006).

Mater. Res. Soc. Symp. Proc. Vol. 1181 © 2009 Materials Research Society   1181-DD02-04

C. Smith[1], M. Pugh[2], H. Martin[2], R. Hill[2], B. James[2],
S. Budak[2], K. Heidary[2], C. Muntele[1], D. ILA[1]

1-Center for Irradiation of Materials, Alabama A&M University, Normal, AL USA
2-Department of Electrical Engineering, Alabama A&M University, Normal, AL USA

Abstract

Effective thermoelectric materials have a low thermal conductivity and a high electrical conductivity. The performance of the thermoelectric materials and devices is shown by a dimensionless figure of merit, $ZT = S^2\sigma/K_{TC}$, σ is the electrical conductivity T/K_{TC}, where S is the Seebeck coefficient, T is the absolute temperature and K_{TC} is the thermal conductivity. In this study we have prepared the thermoelectric generator device of Si/Si+Ge multi-layer superlattice films using electron beam physical vapor deposition (EB-PVD). 5 MeV Si ion bombardment was performed in the multi-layer superlattice thin films to decrease the cross plane thermal conductivity, increase the cross plane Seebeck coefficient and cross plane electrical conductivity.

Keywords: Ion bombardment, thermoelectric properties, multi-nanolayers, Figure of merit.
*Corresponding author: C. Smith; Tel.: 256-372-5875; Fax: 256-372-5867;
Email: cydale@cim.aamu.edu

M. Pugh, R. Hill, B. James, H. Martin are 2008-2009 Senior Design Project Students in the Department of Electrical Engineering in Alabama A&M University.

1. Introduction

The performance of a thermoelectric device is quantified by the dimensionless figure of merit $ZT=S^2\sigma T/K_{TC}$. Our aim is to obtain high ZT values by increasing the Seebeck coefficient S and the electrical conductivity σ, and reducing the thermal conductivity KK_{TC} by bombarding the structure with MeV Si ions. Ion bombardment induces the formation of quantum dots of Si and Ge. In addition to, the quantum well confinement of phonon transmission due to Bragg reflection at lattice interfaces [1,2] the defects and disorder in the lattice caused by ion bombardment and the grain boundaries of these nanoscale clusters increase phonon scattering and increase the chance of an inelastic interaction and phonon annihilation. All these effects inhibit heat transport perpendicular to the layer planes [4–7]. Thus, cross plane thermal conductivity will decrease. These quantum dot layers also increase the Seebeck coefficient and electric conductivity owing to the increase of the electronic density of states produced by the one dimensional periodic potential. We have already studied the improvement of thermoelectric properties for 10–

50 nm multilayers [8]. In this study we report on the growth of Si+Ge multi-layer super-lattice films using co-electron beam physical vapor deposition (EB-PVD). Si and Ge materials were placed in separate electron guns and the shutters were manipulated to deposit the desired materials at the desired thickness. The deposition was followed by a 5 MeV Si ion bombardment at various fluences.

2. Sample preparation and characterization

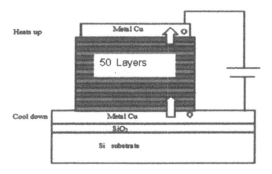

Fig.1. Diagram of electrical conductivity measurements.

Fig. 1 shows the multilayer preparation geometry and the diagram for the electrical conductivity and Seebeck measurement set-up. Si/Si+Ge multilayer thin film thermoelectric (TE) devices were made at the AAMU Center for Irradiation of Materials. These thin films constitute a periodic quantum well structure consisting of 50 alternating layers of total thickness of 300nm. The multilayers were prepared by electron beam physical vapor deposition (EB-PVD). The process pressure $2 \times 10-5$ Torr was maintained throughout the deposition. The multilayer films were sequentially deposited on a Si substrate that was coated with a SiO_2 insulation layer and a metal (Cu) contact layer to form a multilayer. A quartz crystal monitor (QCM) was used to monitor deposition rate and final thickness. For each Si layer the relative rate of deposition was 10 Hz/s from a single e-beam evaporator. The thin film was grown on a carbon substrate for Rutherford backscattering spectroscopy (RBS) analysis. Post 5 MeV Si ion beam process was performed using the AAMU Pelletron accelerator. (SRIM-2008) Stopping Range Ions in

Matter simulation shows that 5 MeV Si ions pass through the multilayer film and terminate deep in the substrate.

The electrical conductivity was measured by the 4-probe contact system and the thermal conductivity was measured by the 3ω technique. The electrical conductivity, thermal conductivity and Seebeck coefficient measurements have been performed at room temperature. In order to make nano clusters in the layers, 5 MeV Si ion bombardments were performed with the Pelletron ion beam accelerator at the Alabama A&M University Center for Irradiation of Materials (AAMU-CIM).

The energy of the bombarding Si ions was chosen by the SRIM-2008 simulation software (SRIM-2008). The fluences used for the bombardment were $1x10^{12} ions / cm^2$, $5x10^{12} ions / cm^2$ and $1x10^{13} ions / cm^2$. Rutherford Backscattering Spectrometry (RBS) was performed using 2.1 MeV He$^+$ ions with the particle detector placed at 170 degrees from the incident beam to monitor the film thickness and stoichiometry before and after 5 MeV Si ion bombardments [9].

3. Thermoelectric measurements and results

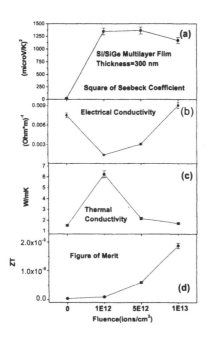

We used the 3ω technique to measure the cross plane thermal conductivity of the thin film samples. The experimental setup has been previously described [3,4,8]. A narrow Ag strip is deposited onto the films providing a heater with a resistance value of about 200 Ω. Originally we got negative Seebeck values for this sample. The negative value of the observed Seebeck coefficient indicates that our samples are n type. For cross plane Seebeck coefficient measurements, the samples are deposited between two thermally and electrically conductive metal (Cu) layers as shown in Fig. 1. The Square of Seebeck coefficients of the multilayer samples before and after 5 MeV Si ion bombardments are compared in Fig. 2a. The electrical conductivities of Si/ Si + Ge multilayer thin films before and after bombardment by 5 MeV Si ions are shown in Fig. 2b. We used a digital electronic bridge with four probe contact system to measure the cross plane electrical conductivity of the thin film samples. We assumed that the Schottky junction barrier between Cu and the semiconductors are negligible and that the resistance of the Cu surface electrodes is negligible. Fig. 2c indicates that cross plane thermal conductivity of Si/ Si + Ge multilayer thin films before and after bombardment by 5 MeV Si ions. As seen from fig. 2c, the thermal conductivity value decreases with increasing ion fluence except for the value at the fluence of $1x10^{12} ions/cm^2$. Finally, we have calculated the figure of merit ZT from the definition ZT=$S^2\sigma$T/K before and after bombardment by 5 MeV Si ions and the results were shown in fig.2d.

4. Discussion and conclusion

In this study we deposited 50 layers as compared to previous studies where we have done on the order of 70 layers. The results here show a slight difference in the seebeck coefficient. This can be attributed to variations in processing parameter leading to variations in the layers of the materials. The increased number of charge carriers due to 5 MeV Si ion bombardment is most dominate feature of this technique for the increasing the electrical conductivity and the seebeck coefficient. The decrease in the thermal conductivity can be related to both the multilayer interface and Si bombardment. We suspect that an increase of thermal conductivity can be accomplished with increasing the homogeneity of the multilayer. By providing more defined interfaces, thus increasing the phonon scattering. Naturally , as the fluence of the ion bombardment increases the more damage is caused in the layers and more importantly the interfaces of the multilayers. We have yet to confirm the presence of nano cluster by microstructure analysis methods such as transmission electron microscopy (TEM). The optimum fluence will yield the greatest number of carriers and phonon scattering sites while minimizing the damage due to post ion beam bombardment.

Acknowledgement
Research sponsored by the Center for Irradiation of Materials, Alabama A&M University and the AAMURI by National Science Foundation EPSCOR CIDEN Grant, and by Department of Electrical Engineering under Nanotechnology Infrastructure Development for Education and Research with the proposal number: 54478-RT-ISP

[1] Xiaofeng Fan, Electron. Lett. 37 (2001) 126.
[2] Xiaofeng Fan, Gehong Zeng, Appl. Phys. Lett. 78 (2001) 1580.
[3] S.M. Lee, D.G. Cahill, Appl. Phys. Lett. 70 (1997) 2957.
[4] D.G. Cahill, M. Katiyar, Phys. Rev., B 50 (1994) 6077.
[5] Rama Venkatasubramanian, Phys. Rev., B 61 (2000) 3091.
[6] J.L. Liu, Phys. Rev., B 67 (2003) 4781.
[7] David G. Cahill, Rev. Sci. Instrum. 61 (1990) 802.
[8] B. Zheng, S. Budak, C. Muntele, Z. Xiao, C. Celaschi, I. Muntele, B. Chhay,
[9]R.L. Zimmerman, L.R. Holland, D. Ila, Materials in Extreme Environments,
Materials Research Society, vol. 929, 2006, p. 81.
[10] L.R. Doolittle, M.O. Thompson, RUMP, Computer Graphics Service
(2002).
[11]S. Budak, S. Guner, R. Minamisawa and D. Ila, Surface and Coatings Technology,
Vol. 203, issues 17-18,2009,P. 2479
[12] B. Zheng, Z. Xiao, B. Chhay, R. Zimmerman, D. Ila , Nuclear Instr. And Methods in
Physics Research B 266 (2008) 73-78

Mater. Res. Soc. Symp. Proc. Vol. 1181 © 2009 Materials Research Society   1181-DD13-05

Fabrication and Characterization of Thermoelectric Generators From SiO₂/SiO₂+Au Nano-Layered Super-Lattices

M. Pugh[1], R. Hill[1], B. James[1], H. Martin[1], C. Smith[2],
S. Budak[1*], K. Heidary[1], C. Muntele[2], D. ILA[2]

1-Department of Electrical Engineering, Alabama A&M University, Normal, AL USA
2-Center for Irradiation of Materials, Alabama A&M University, Normal, AL USA

Abstract

Effective thermoelectric materials have a low thermal conductivity and a high electrical conductivity. The performance of the thermoelectric materials and devices is shown by a dimensionless figure of merit, $ZT = S^2\sigma T/K$, where S is the Seebeck coefficient, σ is the electrical conductivity, T is the absolute temperature and K is the thermal conductivity. In this study we have prepared the thermoelectric generator from the SiO₂/SiO₂+Au multi-layer super-lattice films using the ion beam assisted deposition (IBAD). In order to determine the stoichiometry of the elements of SiO₂ and Au in the grown multilayer films and the thickness of the grown multi-layer films Rutherford Backscattering Spectrometry (RBS) and RUMP simulation software package were used. The 5 MeV Si ion bombardments was performed to make quantum clusters in the multi-layer super-lattice thin films to decrease the cross plane thermal conductivity, increase the cross plane Seebeck coefficient and cross plane electrical conductivity. To characterize the thermoelectric generator before and after Si ion bombardments we have measured the cross-plane Seebeck coefficient, the cross-plane electrical conductivity, and the cross-plane thermal conductivity at different fluences.

Keywords: Ion bombardment, thermoelectric properties, multi-nanolayers, Figure of merit.

*Corresponding author: S. Budak; Tel.: 256-372-5894; Fax: 256-372-5855; Email: satilmis.budak@aamu.edu

M. Pugh, R. Hill, B. James, H. Martin were 2008-2009 Senior Design Project Students in the Department of Electrical Engineering in Alabama A&M University.

1. INTRODUCTION

The theory of thermoelectric power generation and thermoelectric refrigeration was first presented by Altenkirch in 1990 [1]. Thermoelectric materials and devices are being increasingly important due to their applications in thermoelectric power generation and micro electronic cooling devices [2, 3]. Thermoelectric devices do not have moving parts and do not generate greenhouse gases. The efficiency of the thermoelectric devices is limited by the material properties of n-type and p-type semiconductors [4]. Thermoelectric devices have many advantages: lightweight, small, inexpensive, quiet performance and the ability for localized 'spot' cooling [5]. The best thermoelectric materials were succinctly summarized as "phonon-glass electron-crystal" (or PGEC in short), which means that the materials should have a low lattice thermal conductivity as in glass, and high electrical conductivity as in crystals [6]. The efficiency of the thermoelectric devices and materials is determined by the figure of merit ZT [7]. The figure of merit is defined by $ZT = S^2 \sigma T / \kappa$, where **S** is the Seebeck coefficient, σ is the electrical conductivity, **T** is the absolute temperature, and κ is the thermal conductivity [8, 9]. ZT can be increased by increasing **S**, by increasing σ, or by decreasing κ. In this study we report on the growth of $SiO_2/SiO_{2(1-x)}+Au_x$ multi-layer super-lattice films using the ion beam assisted deposition (IBAD), and high energy Si ions bombardment of the films for reducing thermal conductivity and increasing electrical conductivity.

2. EXPERIMENTAL

We have deposited the 50 alternating layers of SiO_2/SiO_2+Au nano-layers films on silicon and silica substrates with the Ion Beam Assisted Deposition (IBAD). The multilayer films were sequentially deposited to have a periodic structure consisting of alternating SiO_2 and SiO_2+Au layers. The two electron-gun evaporators for evaporating the two solids were turned on and off alternately to make multi-layers. The base pressure obtained in IBAD chamber was about 5×10^{-6} torr during the deposition process. The growth rate was monitored by a gold coated Inficon Quartz Crystal Monitor (QCM). The film geometries used for deposition of SiO_2/SiO_2+Au nano-layers films are shown in Fig.1a and 1b for the thermal conductivity and for Seebeck coefficient measurements respectively. The cross plane electrical conductivity was measured by the 4-probe contact system and the cross

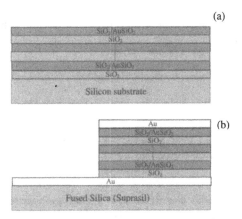

Fig.1. Geometry of sample from the cross-section,
(a) For thermal conductivity, (b) for Seebeck measurements

plane thermal conductivity was measured by the 3ω technique. The electrical conductivity, thermal conductivity and Seebeck coefficient measurements have been performed at room temperature. Detailed information about the 3ω technique may be found in Refs. [10-13]. In order to make nano-clusters in the layers, 5 MeV Si ion

48

bombardments were performed with the Pelletron ion beam accelerator at the Alabama A&M University Center for Irradiation of Materials (AAMU-CIM).

The energy of the bombarding Si ions was chosen by the SRIM simulation software (SRIM). The fluences used for the bombardment were $1x10^{12} ions/cm^2$, $5x10^{12} ions/cm^2$ and $1x10^{13} ions/cm^2$. Rutherford Backscattering Spectrometry (RBS) was performed using 2.1 MeV He$^+$ ions with the particle detector placed at 170 degrees from the incident beam to monitor the film thickness and stoichiometry before and after 5 MeV Si ions bombardments [14, 15].

3. RESULTS AND DISCUSSION

Fig. 2. shows RBS spectrum and RUMP simulation of 50 alternating layers of SiO$_2$/SiO$_2$+Au films on a Glassy Polymeric Carbon (GPC) substrate when the sample is at the normal angle. The same stoichiometry was kept for our 50 alternating layer film deposition on the silicon and silica substrates. RUMP simulation [16] was used both to specify the average thickness of 50 layers for SiO$_2$/SiO$_2$+Au co-deposited layers as 147 nm and the elemental analysis of the thin films.

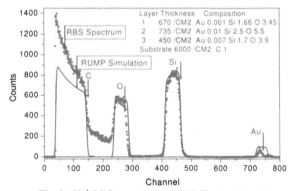

Fig. 2. He$^+$ RBS spectrum and RUMP simulation for SiO$_2$/SiO$_2$+Au nano-layered films on GPC substrate.

Fig. 3 shows the thermoelectric properties of 50 alternating layers of SiO$_2$/SiO$_2$+Au virgin and 5 MeV Si ions bombarded thin films at three different fluences. Fig. 3a. shows the square of the Seebeck coefficient of the thin films. The original Seebeck values are negative that is, we have negative thermo-power and electrons are the main charge carriers. The virgin sample has Seebeck coefficient of −4.80 μV/K and this value decreased from this value until maximum value of -5.65 μV/K at the fluence of $1x10^{13} ions/cm^2$. The observed effects of Si ion bombardment have the opposite characteristics for the electrical and the thermal conductivity values as function of varying fluence, as shown in the Fig. 3b and 3c respectively.

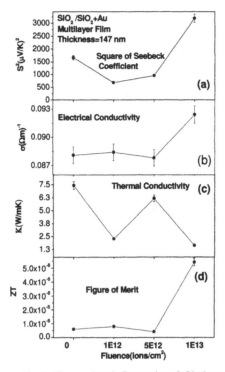

Fig. 3. Thermoelectric Properties of 50 alternating
nanolayers of SiO₂/SiO₂+Au multi-layer films

As seen from fig. 3b, the remarkable increase in the electrical conductivity was observed at the fluence of $1x10^{13} ions/cm^2$. As seen from fig. 3c, the thermal conductivity value decreased when the first ion bombardment was introduced at the fluence of $1x10^{12} ions/cm^2$. After this fluence, the thermal conductivity showed an increase until the fluence of $5x10^{12} ions/cm^2$. The turning point for the thermal conductivity is the fluence of $5x10^{12} ions/cm^2$. After this fluence, the thermal conductivity decreased. Fig. 3d. shows the calculated dimensionless figure of merit, ZT values by applying the equation given in the introduction part. The desired result of ion bombardment on ZT strongly appears at the fluence of $1x10^{13} ions/cm^2$. When we consider the experimental uncertainty, we are able to say that the total effect of bombardment do not differ the ZT values for other fluences except for the fluence of $1x10^{13} ions/cm^2$. Ion bombardment induces the formation of quantum dots of Si and Au in multilayer superlattice thin film system. In addition to, the quantum well confinement of phonon transmission due to Bragg reflection at lattice interfaces [17] the defects and disorder in the lattice caused by ion bombardment and the grain boundaries of these nanoscale clusters increase phonon scattering and increase the chance of an inelastic interaction and phonon annihilation. All these effects inhibit heat transport perpendicular to the layer planes [11]. Thus, cross plane thermal conductivity will decrease. These quantum dot layers also increase the Seebeck coefficient and electric conductivity owing to the increase of the electronic density of states produced by the one dimensional periodic potential when the suitable fluences are introduced.

4. CONCLUSION

We have observed the effects of the ion fluence on the thermoelectric properties of the $SiO_2/Au_xSiO_{2(1-x)}$ alternating layers. The data clearly show that the thermoelectric properties are positively impacted at the suitable fluences. The properties quickly degrade or demonstrate limited response at some fluences as increasing fluence. This behavior suggests ion straggling or damage due to the increasing fluence. We have reported similar study as shown in ref. [2]. We had good results in that study. Our purpose is to improve the figure of merit for this sample system. This study and the previous study gave us a chance to make better future plan to accomplish our goals. In the next studies, we will try thicker film thickness and more Au concentration in the thin film system. Ion fluences will be applied accordingly depending on the sample characterizations in the future studies.

Acknowledgement

Research sponsored by the Center for Irradiation of Materials (CIM), Alabama A&M University (AAMU) and by the AAMURI, by National Science Foundation under NSF-EPSCOR R-II-3 Grant No. EPS-0814103, and by Department of Electrical Engineering under Nanotechnology Infrastructure Development for Education and Research with the proposal number: 54478-RT-ISP.

References
1. Hongxia Xi, Lingai Luo, Gilles Fraisse, Renewable and Sustainable Energy Reviews 11 (2007) 923-936.
2. S. Budak, C. Muntele, B. Zheng, D. Ila, Nuc. Instr. and Meth. B 261 (2007) 1167.
3. S. Budak, S. Guner, C. Muntele, D. Ila, Nuc. Instr. and Meth. B 267 (2009) 1592-1595.
4. Brian C. Scales, Science 295 (2002) 1248.
5. Hongan Ma, Taichao Su, Pinwen Zhu, Jiangang Guo, Xiaopeng Jia, Journal of Alloys and Compounds 454 (2008) 415-418.
6. G. Slack, in: D. M. Rowe (Ed.), CRC Handbook of Thermoelectrics, CRC Press, 1995, p.407.
7. S. Guner, S. Budak, R. A. Minamisawa, C. Muntele, D. Ila, Nuc. Instr. and Meth. B 266 (2008) 1261.
8. B C. -K. Huang, J. R. Lim, J. Herman, M. A. Ryan, J. -P. Fleural, N. V. Myung, Electrochemical Acta 50 (2005) 4371.
9. T.M. Tritt, ed., Recent Trends in Thermoelectrics, in Semiconductors and Semimetals, 71 (2001).

10. L. R. Holland, R. C. Smith, J. Apl. Phys. 37 (1966) 4528.
11. D. G. Cahill, M. Katiyar, J. R. Abelson, Phys. Rev.B 50 (1994) 6077.
12. T. B. Tasciuc, A.R. Kumar, G. Chen, Rev. Sci. Instrum. 72 (2001) 2139.
13. L. Lu, W.Yi, D. L. Zhang, Rev. Sci. Instrum. 72 (2001) 2996.
14. J. F. Ziegler, J. P. Biersack, U. Littmark, The Stopping Range of Ions in solids, Pergamon Press, New York, 1985.
15. W. K. Chu, J. W. Mayer, M. -A. Nicolet, Backscattering Spectrometry, Academic Press, New York, 1978.
16. L. R. Doolittle, M. O. Thompson, RUMP, Computer Graphics Service, 2002.
17. Xiaofeng Fan, Electron. Lett. 37 (2001) 126.

Mater. Res. Soc. Symp. Proc. Vol. 1181 © 2009 Materials Research Society   1181-DD13-10

Band gap engineering of nano scale AlGaN epitaxial layers by Swift Heavy Ion irradiation

N Sathish[1], G Devaraju[1], N Srinivasa Rao[1], A P Pathak[*1], A Turos[2], S A Khan[3], D K Avasthi[3], E Trave[4] and P. Mazzo ldi[4]

[1]School of Physics, University of Hyderabad, Hyderabad – 500 046,India
[2]Institute of Electronic Materials Technology, 01-919 Warsaw, ul. Warsaw, Poland
[3]Inter-University Accelerator Centre, Aruna Asaf Ali Marg, New Delhi 110 067, India
[4] Dip. Fisica "G.Galilei", Università di Padova, via Marzolo 8, 35131 Padova, Italy

ABSTRACT

Epitaxial AlGaN/GaN layers grown by MBE on SiC substrates were irradiated with 150 MeV Ag ions at a fluence of 5×10^{12} ions/cm^2. AlGaN/GaN Multi Quantum Wells (MQWs) were grown on Sapphire substrate by Metal Organic Chemical Vapour Deposition (MOCVD) and irradiated with 200 MeV Au^{8+} ions at a fluence of 5×10^{11} ions/cm^2. These samples were used to study the effects of Swift Heavy Ions (SHI) on optical properties of AlGaN/GaN based nano structures. Rutherford Back Scattering (RBS) /Channelling measurements were carried out at off normal axis on irradiated and unirradiated samples to extract strain. In as grown samples, AlGaN layer is partially relaxed with a small compressive strain. After irradiation this compressive strain increases by 0.22% in AlGaN layer. Incident ion energy dependence of dechannelling parameter shows $E^{1/2}$ dependence, which corresponds to the dislocations. Defect densities were calculated from the $E^{1/2}$ graph. As a result of irradiation defect density increased on both GaN and AlGaN layer. Optical properties of AlGaN/GaN MQWs before and after irradiation have been analyzed using PL. In this study, we present some new results concerning high-energy irradiation on AlGaN/GaN heterostructures and MQWs characterized by RBS/Channelling and Photo Luminescence (PL).

---

*Corresponding author. Tel.: +91 40 23010181 / 23134316 Fax: +91 40 23010181 / 23010227.
E-mail address: appsp@uohyd.ernet.in   &   anandp5@yahoo.com (A.P. Pathak).

INTRODUCTION

III- Nitride semiconductors are a novel class of materials for optoelectronics & high power, high temperature device applications and are used in wide range of optical devices like Blue Light Emitting Diode (LED) to violet laser diode (LD) & UV photo detectors and High Electron Mobility Transistors (HEMTs). These materials are the best candidates for microwave electronics for base stations of cell phones. AlGaN/GaN MQWs also have numerous optoelectronic applications including semiconductor photodiodes. Quantum well interdiffusion technology has become increasingly important in the drive towards fabrication of photonic integrated circuits due to its versatile band gap tuning process [1,2]. SHI beam deposits energy via Electronic energy loss mechanism and create defects into the quantum well active region, which allows atomic diffusion to take place between the quantum well and barrier materials. Recently we have demonstrated strain modification in lattice-matched heterostructures using SHI and extensively strain relaxed MQWs have also been studied [3]. The possibility of material reconstruction and interface smoothening has been demonstrated using SHI beams. In a detailed study, S O Kucheyev el.al [4] have studied the MeV ion impact on GaN, AlN, AlGaN and InGaNs. It has been observed that AlN is more radiation resistant than GaN. The process of dynamic annealing is stronger in AlN because of its larger Al-N bond strength. Here, we present some new results of high-energy irradiated AlGaN/GaN systems. Subsequently, RBS/Channelling and PL measurements were carried on all as grown and irradiated samples.

EXPERIMENTAL DETAILS

Multi quantum wells were grown on C-sapphire by Metal Organic Chemical Vapour Deposition (MOCVD) system equipped with a horizontal quartz reactor at ITME Warsaw, Poland. Trimethyl Gallium (TMGa), Trimethyl Aluminum (TMAl) and ammonium were used as Ga, Al and N_2 sources in addition to hydrogen as a carrier gas. AlN was deposited as a nucleation layer of thickness 20 nm at a temperature of 550°C and pressure of 70 mbar, on sapphire substrate. Subsequently, a GaN nucleation (buffer) layer of thickness 200 nm was deposited at temperature of 1140°C and pressure of 200 mbar. This was followed by deposition of 15 periods of multi quantum wells, each of thickness 13 nm, at a temperature of 1140°C and pressure of 70 mbar. Besides these Al0.49Ga0.51N/GaN MQWs, we have also investigated $Al_{0.2}Ga_{0.8}N$ / GaN heterostructures grown on semi-insulating (SI) 4H-SiC substrates. Such grown structures were irradiated with 200 MeV Au^{8+}, 150MeV Ag ions at a fluence of $5x10^{11}$ ions/cm^2 and $5x10^{12}$ ions/cm^2 respectively. Irradiation was performed at room temperature using NEC 15 MV pelletron accelerator at Inter University Accelerator Centre, New Delhi. A low beam current (0.5–2 pnA) was maintained to avoid heating of the samples and the samples were oriented at an angle of about 5-7^0 with respect to the beam axis to minimize channelling during irradiation. According to SRIM 2003 calculations, the projected ranges for 150 MeV Ag, 200 MeV Au are larger than the sample thickness. Hence it makes uniform modification throughout the layer of interest. For GaN, nuclear energy loss is 9 $x10^{-2}$ keV/nm with Sn / S_e ratio as $3.6x10^{-3}$. Therefore these ions are used to avoid high Sn/Se ratio and possible amorphisation of the sample. Photo luminescence studies were carried out on as grown and irradiated MQW samples at room temperature (295K). PL was excited with Xe UV lamp (265nm) and detected with HAMAMATSU N2-cooled detector. RBS/Channeling measurements from as grown and irradiated samples yield strain. Energy dependence of dechannelling parameter has also been analyzed. Dependence of incident ion energy to the dechannelling parameter will give us the quantitative and qualitative information about the defects present in the crystal. The backscattered yield is collected using the detector (mounted at a scattering angle of ~170°) and the nuclear instrumentation modules. <001> axial channelling was carried out for dechannelling analysis varying the incident energies between 2 and 4.1 MeV. The dechannelling parameter is calculated from the normalized back scattering yield to see the energy dependence for defect analysis.

RESULTS AND DISCUSSION

Photoluminescence of unirradiated MQW sample shows a weak A and I2 transitions from GaN bulk crystal because most of the light will be absorbed in the multilayer. Luminescence around ~3.7 eV comes from the QWs, which implies that GaN layers are confined between AlGaN layers. Moreover, we did not observe any yellow luminescence ~ 2.1 eV for this unirradiated sample. Irradiated MQWs show strong yellow luminescence at ~2.1 eV (see fig 1(a)), which is generally observed due to the deep levels like Ga vacancy or oxygen related defects or dislocation related luminescence. In the irradiated sample, exciton structures have vanished and one exciton transition I2 (~ 3.45 eV) is observed , which may be due to the bound exciton to neutral donors commonly observed in undoped n-GaN. Quantum Well luminescence intensity has been increased by one order which shows that the confinement effects are enhanced in irradiated samples [5].

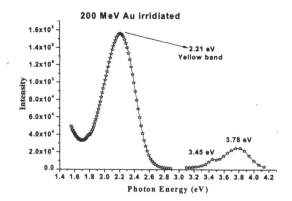

Fig.1 (a)

Our previous studies to get insight into the details of the mechanism of optical properties on the effects of SHI in AlGaN/GaN has been carried out using RBS/C, High Resolution X-Ray Diffraction (HRXRD) and Atomic Force Microscope (AFM) [6].

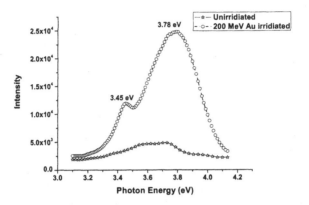

Fig.1 (b)

Fig.1: (a) PL spectra of irradiated sample recorded at RT, Shows Yellows luminescence (b) GaN band edge luminescence of Unirradiated and irradiated samples, shows after irradiation GaN luminescence intensity increased by one order.

Here we present only the strain and defect density results obtained from RBS/Channelling measurements. In un irradiated sample, the layers are partially relaxed and AlGaN layer is still under a compressive strain of $\varepsilon_t = -0.55$ % (See Fig 2). Dechanneling measurements show $E^{0.5}$ behaviour and the defect density is found to be 2.1×10^8 cm^{-2} for GaN layer and 1.9×10^8 cm^{-2} for AlGaN layer. These values are close to each other. This may be due to the fact that the calculated defect density both within the layer as well as the region just below the AlGaN layer.

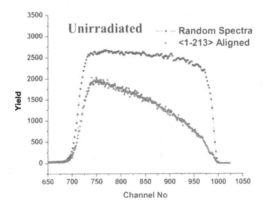

Fig.2 (a)

56

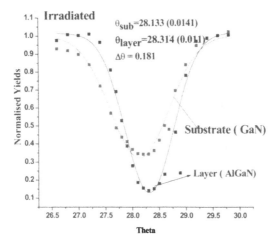

Fig.2 (b)

Fig.2. (a) RBS/ Channelling spectra of unirradiated AlGaN/GaN heterostructure sample (b) Angular scans of Irradiated sample for the strain measurement on the off normal axis.

As seen from the calculated value in fig 2, irradiation has resulted in dynamic annealing which has introduced a compressive strain in AlGaN layer, which is given as $\varepsilon_t = -0.76$. The increase in PL luminescence intensity of AlGaN layers is due to the induced strain in the layers. Defect density of GaN layer is 3.6×10^8 cm^{-2} and AlGaN layer is 3.2×10^8 cm^{-2}. We notice that there is an increase in the defect densities after irradiation, in both the layers. Here we speculate that the origin of yellow luminescence is from the irradiation-induced dislocations and point defects. However, the increase in defect density in AlGaN layers is lesser than GaN, which may be due to induced strain in the epilayer.

CONCLUSIONS

Swift heavy ion irradiation effects on relaxed AlGaN/GaN heterostructures and MQWs have been analysed using RBS/Channelling and PL. This study shows that SHI increases the confinement effects in the MQWs and intensity of the active layer of the MQWs luminescence is increased by one order. This may be due to the induced strain in GaN and AlGaN layers. Channelling strain measurement shows that unirradiated sample has a small residual compressive strain, and it increases after irradiation. At the same time the defect density in both these AlGaN and GaN layers also increases as seen from the calculated values for irradiated samples and the unwanted yellow luminescence has also been introduced by the SHI. It may be due to the defect luminescence from the point defects or the induced dislocations in GaN bulk epitaxial layers.

The projectile energy transfer depends on the material properties. In III-Nitrides, the process of dynamic annealing is very strong because of their high bond strength. Al-N bond is stronger than Ga-N, so Al-N recombines faster and complete epitaxial reconstruction of AlGaN layer has been observed. Consequently, we observed more threading dislocations in both the layers. Swift heavy ion induces more defects in GaN than in AlGaN layer by the process of dynamic annealing. This same process of dynamic annealing also induces more strain in AlGaN layer, which increases the luminescence intensity of quantum well. This opens new area of investigations in material reconstruction of AlN based III-Nitrides and more detailed studies have to be carried out to optimise the ion beam parameters.

ACKNOWLEDGEMENTS

N S would like to thank SSPL, CARS and UGC-CAS for fellowship. G D would like to thank Center for Nano Technology for fellowship through DST-Nano project sanctioned to APP. This work has been partly supported by Indo-Polish Scientific and Technological Cooperation Programme

REFERENCES

1. S C Jain, M. Willander, J. Narayan and R. Van Overstraeten, .J Appl Phys. **87**, 965 (2000).
2. S J Pearton, C R Abernathy, M E Overberg, G T Thaler, A H Onstine, B P Gila, F Ren, B Lou and J Kim, Materials today. June 2002.
3. S Dhamodaran, A P Pathak, A Turos, and B M Arora, Nucl. Inst. and Meth. B **266**, 1908 (2008).
4. S O Kucheyev, J S Williams and C Jagadish, Vaccum, **73**, 93 (2004).
5. B. Monemar, P. P. Paskov, T. Paskova, J. P. Bergman, G. Pozina, W. M. Chen, P. N. Hai, I. A. Buyanova, H. Amano, I. Akasaki, Mat. Sci. and Eng B **93**, 112 (2002).
6. N Sathish, S Dhamodaran, A P Pathak, B Sundravel, K G M Nair, S A Khan,D K Avasthi, M. Bazzan E. Trave, P. Mazzoldi and D. Scott Katzer, Nucl. Inst Meth B. (Communicated)

Nanostructure Formation and
Fabrication of 3D Structures

Mater. Res. Soc. Symp. Proc. Vol. 1181 © 2009 Materials Research Society   1181-DD03-01

Fabrication of micro-fluid-channel structures by focused ion beam techniques

Junichi Yanagisawa[1], Hiroaki Kobayashi[2], Kakunen Koreyama[2], and Yoichi Akasaka[2]
[1]School of Engineering, The University of Shiga Prefecture,
2500 Hassaka-cho, Hikone, Shiga 522-8533, Japan
[2]Graduate School of Engineering Science, Osaka University,
1-3 Machikaneyama-cho, Toyonaka, Osaka 560-8531, Japan

ABSTRACT

One of the new applications of focused Ga ion beam (Ga FIB) techniques in the fabrication of micro-fluid-channels on plate glass was demonstrated. After discussing the features of the FIB-etched patterns, narrow or Y-shaped channels were fabricated by FIB etching on a patterned plate glass prepared by photolithography and wet etching. Micro-fluid devices were then constructed using a polydimethylsiloxane (PDMS) sheet and silicone rubber tubes, and the water (or ink) flow in the devices was observed under a microscope using a syringe pump. Although no discussion based on fluid mechanics has been carried out at present, the present results indicate the possibility of applying FIB techniques to fabricate micro-fluid devices that can be used in bio- and/or chemical-related fields.

INTRODUCTION

Among many microfabrication techniques developed so far, mainly in the semiconductor research and manufacture fields, photolithography using masks with fixed patterns has potential for the mass production of devices. This technique is now widely used in other fields such as bio- and chemical-chip fabrication on plate glass. In bio- and chemical chips, micro-fluid channels are key structures. By photolithography and etching, the depth of the micro-fluid channels can be controlled by adjusting the etching time, therefore, the depth of patterns produced during one process is almost uniform. In other words, it is difficult to fabricate micro-fluid channels with partially different depths by conventional lithography techniques. In addition, only one chip with different kinds of patterned microchannel structures is desired in research and development stages. For such usages, photolithography may not be the best production process.

Using a focused ion beam (FIB), the fabrication of microstructures in any pattern with any depth can be accomplished without using any masks. In addition, novel three-dimensional structures can also be formed by the FIB process [1]. Although the process time of FIB etching is rather long (depending on the patterning area desired), it is expected to be used in research and development fields in which only one sample chip is required. Wilhelmi *et al.* reported the formation of Y-shaped trench structures on Si wafer by FIB milling [2]. However, many bio- and chemical chips are formed on plate glass at present. Therefore, to apply the FIB process in producing test bio- and chemical chips, the formation of microchannels by FIB should be performed on plate glass.

In the present work, the formation of microchannels on plate glass by the FIB process is performed. Although no discussion based on fluid mechanics has been conducted at present, several examples of the actual flow in the FIB-fabricated channels are shown below.

EXPERIMENTAL DETAILS

In terms of process time, the formation of the entire microchannel by only the FIB process is unrealistic. Therefore, it is convenient to fabricate common structures, such as inlet and outlet channel structures, as well as guiding channels to the area in which FIB patterning is to be performed, on plate glass by the conventional lithography and etching processes. The plate glass used in this study was borosilicate plate glass (Matsunami Glass) with a size of 38 x 26 mm^2 and a thickness of 1 mm.

After cleaning the glass surface using organic solvent (acetone and 2-propanol) and a mixture of sulfuric acid and oxygenated water, followed by rinsing, thin Cr (50 nm) and thick Au (totally 1000 nm) films were deposited on the glass surface using an electron beam evaporator. For details on this process, refer to the reports by Iliescu et al. [3, 4]. To fabricate common structures, the 1st photolithography was performed using a mask, as shown in figure 1 (a). The pattern has three circles 1 mm in diameter for the inlet and outlet, and 200- or 100-μm-wide guiding channels from each circle to the 2 mm x 2 mm square region, where the 2nd etching is to be performed. After patterning etching of the Au and Cr films using a water solution of iodine and potassium iodine and of ceric ammonium nitrate, respectively, the glass surface was etched using hydrofluoric acid to the channel depth of about 70 μm. Because it is difficult for liquid to flow in narrow and shallow channels with long lengths in the center region, rather wide and deep channels were formed as bypasses. On the other hand, a typical FIB etching pattern is several μm in width and depth, which is far smaller than the guiding channel formed. Therefore, we performed the 2nd photolithography in the 2 mm x 2 mm region using a mask, as shown in figure 1 (b); then shallow etching (less than 10 μm) was performed. After these processes, Au film was removed and Cr film was left on the initial glass surface; it can be used as the electrode to counter charges during the FIB irradiation. Figure 1 (c) shows the scanning electron microscope (SEM) image of an example of the final "blank" pattern formed on the plate glass. The FIB etching can be performed to connect each end point of the channels with a desired pattern. In the present study, a 100 keV focused Ga ion beam (JIBL-104A, JEOL) was used at fluencies from 2.5 x 10^{18} to 1 x 10^{19} cm^{-2}, repeating fast scans many times. After the FIB process, the Cr layer was removed.

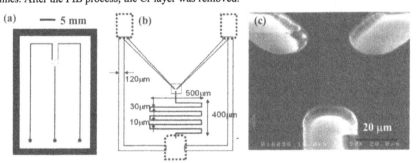

Figure 1. Examples of (a) a large mask pattern which determines inlet and outlet positions and wide channels with a width of 100 or 200 μm to the confluent position, and (b) a small mask pattern to reduce the confluent area for FIB patterning with bypasses. (c) Example of transferred blank pattern on a plate glass of the confluent position (viewing angle: 45°).

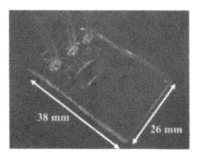

Figure 2. Overview of completed micro-fluid device used in the present study, formed on plate glass with a PDMS cap sheet and inlet and outlet tubes.

To produce micro-fluid devices, polydimethylsiloxane (PDMS) sheets were made using SILPOT 184 (DOW CORNING TORAY) and through-holes were punched in the sheet in accordance with the position of the inlet and outlet formed on the plate glass. After the surfaces of the plate glass and the PDMS sheet were cleaned using oxygen plasma, the surfaces were attached to each other, then the two materials were strongly bonded. Then the silicone rubber tubes were inserted and fixed at the holes using PDMS. A photograph of the fabricated device is shown in figure 2.

In the examination of the liquid flow in the microchannels fabricated, a programmable syringe pump (PHD2000, HARVARD) was used. A typical pumping speed was 100 μl/s. Water or ink (black and red) was injected from the inlet and the flow in the microchannel was observed under an optical microscope.

RESULTS AND DISCUSSION

Observation of FIB etched patterns

To investigate the potential of FIB etching on plate glass, several Y-shaped channels were fabricated on the patterned sample as shown in figure 1 (c). Figure 3 (a) shows the planned FIB patterns used. The FIB fluence on the narrow (2 μm) channels was 2.5×10^{18} cm^{-2}, and that on wide (5 and 10 μm) channels was 5×10^{18} cm^{-2}. Figures 3 (b) – 3 (d) show SEM images of the channels fabricated on plate glass. The fluence on the overlapping region is, therefore, 7.5×10^{18} cm^{-2}.

First of all, it was found that the plate glass was etched finely without the charging effect during FIB irradiation. This might be because of the Cr layer left on the glass surface. The etched surface was rather smooth, and the sidewalls were very steep. The etched depth was controlled well by the FIB fluence. As a result, it is shown that microchannels of different depths, as well as widths, can be fabricated in one FIB process; this is difficult to accomplish by the conventional photolithography and etching. This result indicates the potential of the FIB process in fabricating bio- and chemical chips.

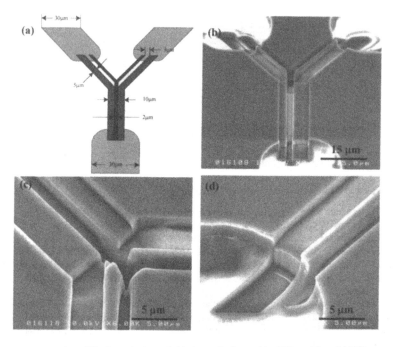

Figure 3. Examples of Y-shaped micro-fluid channels formed by FIB etching. (a) FIB patterns used. FIB fluence on the narrow (2 μm) channels was 2.5 x 10^{18} cm^{-2}, and that on wide (5 and 10 μm) channels was 5 x 10^{18} cm^{-2}. (b)-(d) SEM images of the fabricated channels on plate glass (viewing angle: 45°).

Liquid flow in a narrow channel

As shown in figure 4 (a), a straight-through blank structure (length of 25 μm) was formed at the middle part in a microchannel with the width of 35 μm on the plate glass. Then the Ga FIB was irradiated repeatedly on the blank region, forming a narrow channel with a width of 10 μm through the blank region, as shown in figure 4 (b). After the plate glass was capped with the PDMS sheet, the water flow in the narrow channel region was observed. The observed water flow in the channel is shown in figure 5 (in the page after next).

The flow of water, which was shown as the area of higher brightness in the channel, was from the right to the left in the microchannel. The time interval between adjacent pictures was 1/30 sec. The velocity of the water in the channel after passing the narrow channel formed by FIB etching (positioned at the center part, as shown in figure 4) was, therefore, estimated to be about 13.5 mm/s, which is far slower than the velocity before passing the narrow channel.

From this result, it is found that the water flow can be controlled by the microstructures formed by the FIB process.

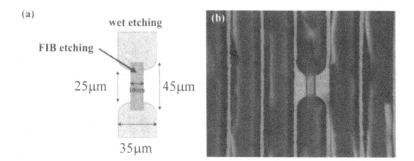

(a) wet etching

FIB etching

25μm | 10μm | 45μm

35μm

(b)

Figure 4. (a) FIB pattern used to connect microchannels formed by wet etching. (b) Microscope image of the fabricated channel.

Liquid flow in Y-shaped channels

In this section, a mixture of two kinds of liquid flow was observed using several Y-shaped microchannels, as shown in figure 6. Figure 6 (a) shows a wide Y-shaped channel formed by conventional photolithography and wet etching (no FIB process was used). Using the blank pattern shown in figure 1 (c), symmetric and asymmetric Y-shaped channels were formed by FIB etching, as shown in figures 6 (b) and 6 (c), respectively. After the capping using the PDMS sheet on the plate glass surface, black and red ink was made to flow from the two inlet channels (upper two channels shown in figures 6 (a) – 6 (c)). The results are shown in figures 7 – 9, where the two channels on the right are inlet channels.

When using the wide Y-shaped channel shown in figure 6 (a), backflow of black ink into another inlet channel was observed, as shown in figure 7. This might be because that the pressure at the confluence cannot be controlled owing to the large width of the channels. Therefore, the narrow and shallow inlet channels, as shown in figures 6 (b) and 6(c), were used next. The results are shown in figures 8 and 9.

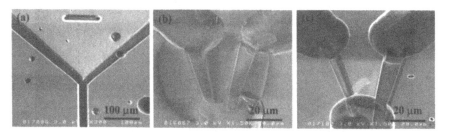

Figure 6. SEM images of Y- shaped microchannels formed on plate glass. (a) Wide channels formed by wet etching only. (b) and (c) Symmetric and asymmetric narrow channels formed by FIB etching using blank patterns shown in figure 1 (c) (viewing angle: 45°).

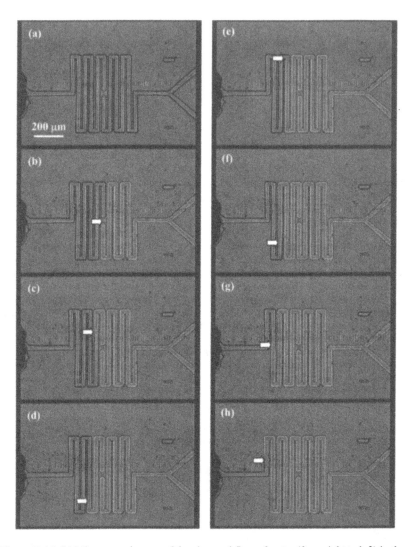

Figure 5. (a)-(h) Microscope images of the observed flow of water (from right to left) in the microchannel. The time interval between adjacent pictures was 1/30 sec. The front of the water flow in the microchannel was indicated by the white arrow. The velocity of the water in the channel after passing through the narrow channel formed by FIB etching (positioned at the center part, as shown in figure 4) was estimated to be about 13.5 mm/s, which was far slower than those before passing the narrow channel.

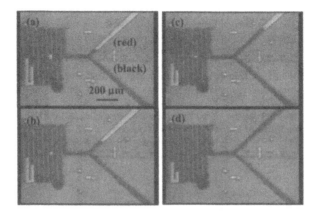

Figure 7. (a)-(d) Observed backflow of black ink into another inlet channel filled with diluted red ink. The inlet channels were two channels on the right, and the outlet channel was that on the left after the confluence.

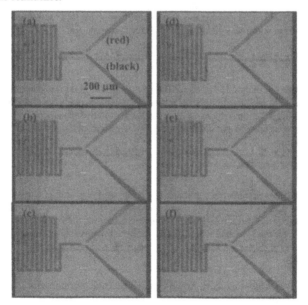

Figure 8. (a)-(f) Observed flow of red and black ink using the symmetric narrow Y-shaped channel shown in figure 6 (b). The time interval between adjacent pictures was 1/30 sec. After passing the confluence, two-phase flow of almost equal widths was observed for a rather long distance.

Figure 8 shows the observed flow of red and black ink using a symmetric narrow Y-shaped channel. The time interval between adjacent pictures was 1/30 sec. No backflow was observed in this structure. This indicates that the pressure can be controlled by introducing the narrow channels. After passing the confluence, two-phase flow with almost equal widths was observed for a rather long distance.

To control the flow volume in the channel after the confluence point, the channel width was changed asymmetrically, as shown in figure 6 (c). The flow volume was clearly changed, as shown in figure 9.

These results indicate the possibility of controlling the liquid flow in the microchannels by introducing small structures formed by the FIB process.

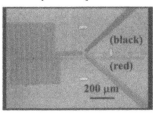

Figure 9. Observed flow of red and black ink using the asymmetric narrow Y-shaped channel shown in figure 6 (c). After passing the confluence, two-phase flow with a ratio of almost 1(black):12(red) in width was observed for a rather long distance.

CONCLUSIONS

Fabrication of microchannels by FIB etching on plate glass was demonstrated. From the observation of the liquid flow in the microchannels, it can be concluded that FIB is a potential tool for fabricating functional micro-fluid devices. Quantitative discussion based on fluid mechanics will be needed to explain the liquid flow observed in the present experiment, and is a task for the future.

ACKNOWLEDGMENTS

The authors would also like to thank K. Gamo, Y. Yuba, and A. Sakai for useful discussions and K. Kawasaki for his technical assistance. This work was partially supported by the 21st Century COE Program (G18) of the Japan Society for the Promotion of Science. One of the authors (J.Y.) is also grateful to the Murata Science Foundation for financial support.

REFERENCES

1. S. Matsui, T. Kaito, J. Fujita, M. Komuro, K. Kanda, and Y. Haruyama, *J. Vac. Sci. Technol.* **B18**, 3181 (2000).
2. O. Wilhelmi, S. Reyntjens, C. Mitterbauer, L. Roussel, D. J. Stokes, and D. H. W. Hubert, *Jpn. J. Appl. Phys.* **47**, 5010 (2008).
3. C. Iliescu, J. M. Miao, and F. E. H. Tay, *Sensors and Actuators* **A 117**, 286 (2005).
4. C. Iliescu, F. E. H. Yay, and J. Miao, *Sensors and Actuators* **A 133**, 395 (2007).

Mater. Res. Soc. Symp. Proc. Vol. 1181 © 2009 Materials Research Society   1181-DD03-04

Christoph Ebm[1] and Gerhard Hobler[2]

[1] IMS Nanofabrication AG, Schreygasse 3, A-1020 Vienna, Austria
[2] Institute of Solid State Electronics, Vienna University of Technology, A-1040 Vienna, Austria

ABSTRACT

Ion-beam induced etching and deposition rates are proportional to the flux of recoils reaching the surface. Based on this finding we propose an improved algorithm for etching and deposition simulations. In this algorithm the recoil flux at each point on the surface is calculated by summing up the recoil fluxes originating from ions impinging on any other surface point. The latter are determined by interpolation in tables calculated by binary collision simulations. For concave surfaces a correction to this algorithm is proposed. Fluxes calculated by this model are in good agreement with binary collision simulations of collision cascades in the same 2-d structure. Consistent with experimental findings, the model predicts that deposited pillars are broader than the ion beam, while etched trenches do not show such broadening. The pillar broadening is related to the lateral straggling of the recoils.

INTRODUCTION

Sputtering and gas-assisted etching and deposition by focused ion beams are established techniques for direct fabrication of nanostructures [1,2]. They are presently used, e.g., for integrated circuits modification, rapid prototyping of photonic structures and photomask repair. With the development of tools providing massively parallel focused ion beams [3,4] higher throughput will be possible, and new applications will emerge like the fabrication of leading-edge photomasks or direct patterning for nanotechnology applications [5] including the fabrication of master stamps for nanoimprint lithography (NIL) [6].

Topography simulation of the gas-assisted etching/deposition process potentially is a valuable tool for investigating the effects of process variations and, in general, to understand the phenomena involved. In particular, nanometer sized pillars formed by ion beam induced gas-assisted deposition have been found to be broader than the ion beam [7], which sets a lower limit to the achievable sizes.

There are two competing theories of the mechanism responsible for the dissociation of the precursor molecules, by secondary electrons or by recoils. In [8] good evidence is given that the deposition yield is proportional to the sputtering yield and thus recoils yield, while no correlation with the yield of secondary electrons has been found. We therefore prefer the recoils theory and explain pillar broadening by the range and lateral spread of the recoils inside the pillar. Simulation together with appropriate experiments could confirm this model.

Existing codes might be extended from sputtering simulation to deposition/etching simulation by assuming that the rate of deposition/etching is proportional to the flux of sputtered atoms. This is justified since the rate of both phenomena is proportional to the number of recoils reaching the surface. However, the codes are based on a local model of sputtering [9], i.e., sputtering from a given point at the surface is calculated from the energy, amount, and angle of incidence of the ions locally at the point where they impinge. This approach is inaccurate if the feature sizes are comparable or smaller than the range of the ions in the substrate [10] and, in particular, pillar broadening cannot be predicted. This is illustrated in Figure 1a, which shows an overlay of recoil cascades caused by the impact of 30 keV Ar ions on a flat target [11] and a deposition simulation with the non-local algorithm described later in this paper. In this example zero width of the ion beam has been assumed. Thus, in a strictly local model the pillar should have zero width as well. In contrast, a width of about 43 nm is observed. It can be seen that many recoils exit the pillar on the side and not on the top. This causes a horizontal growth of the pillar. After some time the deposit is higher than the range of the ions in the substrate. At points that can no longer be reached by recoils the horizontal growth comes to an end. This explains why a deposited pillar has vertical sidewalls and not a Gaussian profile as might be expected from an ion beam with a Gaussian intensity profile.

In Ref. [10] we introduced a non-local algorithm which is in good agreement with sputter flux calculations of 2-dimensional binary collision simulations for planar or convex structures, but has significant problems describing the fluxes at concave surfaces. In this paper we present a refined algorithm that improves the treatment of concave structures. In addition, we investigate the effect of the non-local model on deposition and etching.

SIMULATION

Existing Models

All topography simulations are performed with the IonShaper® software [12,10]. In IonShaper® the surface is described by a sufficient number of points which are moved perpendicular to the average slope of the adjacent line segments. The velocities of the points are calculated from the fluxes of atoms sputtered by the incident ion beam and by ions reflected from other parts of the surface, and from the fluxes of redeposited atoms originating from sputtering at other surface points. Sputtering is treated as a local process. Gas-assisted etching and deposition is assumed to be proportional to the local flux of recoils reaching the surface and the precursor coverage. The latter is close to unity under the conditions studied in this paper.

In order to include non-local effects, tables of the spatial distributions of recoils reaching the surface are used that have been calculated by binary collision simulations using IMSIL [13] assuming a planar surface and a zero-sized beam at various incidence angles α [10]. The recoil flux at a destination point (D) is calculated by summing up the contributions f_r from all possible source points (S) considering the distance d_0 between S and D and the virtual incidence angle α that would be observed if the surface were planar through points S and D, see Figure 1b. While the non-local model increases the computational expense of the IonShaper® simulations, it is still much more efficient than direct binary collision simulations.

Improved Non-Local Recoil-Based Model

The model described above works excellently for convex structures. However, for concave surfaces significant deviations compared to 2-d IMSIL binary collision simulations have been observed [10]. This is because in order to be counted as sputtered at some destination point D when the ion impinges at some source point S (see Figure 1b), recoils have to travel through the target from S to D. In case of a concave surface the target volume available for this travel is reduced, which is not taken into account by the existing model. For the new model we assume that the decrease of sputtering in the concave case can be described in terms of the length d of the contour between S and D. Using binary collision simulations in various concave test structures we found that the amount of recoils reaching the destination point decreases exponentially with respect to the fraction d/d_0. Additionally, it turns out that for grazing incidence (θ and α close to 90°) the relationship is more universal when only the fractions $1-Y_b$ of the ion fluxes that remain in the target are considered, where the ion backscattering yields Y_b have also been obtained from IMSIL simulations. The modification of the flux f_r from the previous model at concave surfaces thus reads

$$f_{r,concave} = f_r \cdot \exp\left[-k\left(\frac{d}{d_0}-1\right)\right] \cdot \frac{1-Y_b[\theta]}{1-Y_b[\alpha]} \tag{1}$$

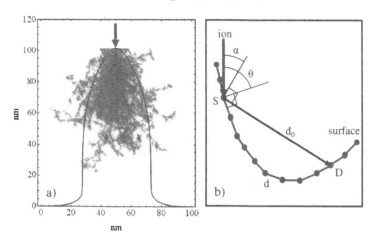

Figure 1: (a) Overlay of 30 keV Ar recoil cascades and a deposition simulation using the non-local algorithm. (b) Definition of the relevant parameters of the non-local recoil-based algorithm implemented in the IonShaper® simulation software.

RESULTS

Recoil Flux Calculations

Figure 2 shows the contour of a microtrench in Si (dotted line) that forms during sputtering in the corner of a bigger trench due to reflection of ions at the sidewall on the left. Using this geometry and a homogeneous distribution of incident 10 keV Ar ions, the flux of sputtered atoms was calculated along the surface by 2-d binary collision simulations using IMSIL and by two different IonShaper simulations. In these simulations the ions and recoils were stopped when leaving the target, i.e. ion reflection and redeposition were not considered.

While these effects significantly affect the shape of the evolving surface, it is advantageous to neglect them in this comparison in order to specifically study the non-local recoil effect and to avoid potential uncertainties stemming from different treatment of reflection and redeposition in the two simulation techniques. The result of the binary collision simulation, which is considered a reliable reference, is shown in Figure 2 by the solid red line. The green, dash-dotted line shows the IonShaper® result using the recoil-based model without special consideration of concave structures [10]. Excellent agreement can be observed everywhere except in the concave region between 33 and 45 nm of the lateral coordinate, where the old IonShaper® model significantly overestimates the primary sputtering flux. As a consequence, the old model would yield a too sharp corner of the microtrench when the effect of ion reflection is added. In contrast, the results obtained with the improved non-local model (dashed, blue line) are in remarkably good agreement with the binary collision simulations over the whole surface.

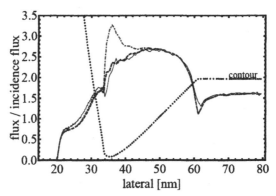

Figure 2: Comparison of recoil flux at the surface for 10 keV Ar on Si. Solid (red) line: emission flux taken from 2-d binary collision simulation as a reference; dash-dotted (green line): emission flux from non-local model without special consideration of concave structures; dashed (blue) line: emission flux from improved non-local algorithm; dotted (black) line: contour for the experiment assuming same scale in vertical and lateral direction.

Etching and Deposition Simulations

In Figure 3 deposition and etching simulation results using the non-local algorithm are shown. 30keV Ar ions, a Si substrate, as well as XeF$_2$ and a hypothetical precursor for Si deposition were used. As indicated in the upper part of the panels the beam profile was assumed to be 12.5nm wide without blur. This beam width was chosen because it corresponds to the beam width of our CHARPAN tool. No blur was assumed in the simulations in order to allow a more straightforward demonstration of the non-local effects.

The deposited pillar in Figure 3a is considerably wider than the beam. Since no blur of the beam was assumed, this can only be due to the range of the recoils ejected on the sides of the pillar. Figure 3b shows the simulation of a trench formed by gas-assisted etching. Contrary to the pillar the trench does not grow wider than the ion beam, because once the trench is as wide as the beam no more ions hit the sidewalls and thus no recoils are produced to activate the precursor.

Recoils are mainly formed at the bottom of the trench. It can be seen from the recoil cascades in Figure 1a that only a few recoils return to the height of the point of incidence of the ion beam or higher. Thus there is no significant etching of the sidewalls due to the finite range of the recoils, although the vertical growth of the trench is accelerated considerably by the etching process.

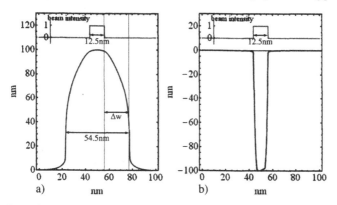

Figure 3: IonShaper® simulation of 30 keV Ar → Si. a) Deposition using the non-local algorithm. b) Etching using the non-local algorithm. The beam intensity profile is indicated in the upper part of the figure panels. Δw is the increase in pillar width on either side of the ion beam.

Relationship between Pillar Broadening and Lateral Straggling of the Recoils

Using our non-local algorithm we have simulated the deposition of pillars for Ar and Xe ions at 10 and 30 keV each. We defined the pillar broadening Δw as the increase in pillar width on either side of the sharp ion beam, i.e. half the difference between pillar and beam width. Among various cascade parameters (range, vertical and lateral straggling of ions and vacancies/interstitials) the best correlation with the beam broadening was found for the lateral straggling σ_{lat} of the vacancies/interstitials. Figure 4 shows a linear relationship. The pillar broadening Δw is about twice as much as the lateral spread. This relationship can be used to estimate the minimum achievable feature size of deposited features for a given precursor, ion and ion energy combination.

CONCLUSIONS

We have proposed a new non-local algorithm for ion-beam induced etching and deposition. With this algorithm it can be explained that nanostructures created by ion-beam induced deposition are considerably wider than the ion-beam, while etched structures do not show an increased width. From simulations for different ion species and energies we found a relationship between the additional width and the lateral straggling of the recoils in the material. This allows for an estimation of the minimal achievable feature size for a given process. Furthermore our model gives an explanation for the perpendicular sidewalls of the deposited pillars.

The present model is based on binary collision simulations and should be reasonably reliable. To further verify the simulations we plan to conduct dedicated experiments.

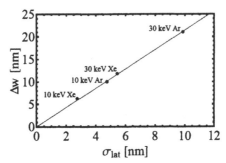

Figure 4: Relationship between change in width of deposits on either side of the beam and the lateral spread of recoils inside the substrate. Data given is for Si deposit.

ACKNOWLEDGMENTS

This work has been partly supported by the European commission through funding of the CHARPAN integrated project and the Austrian Promotion Agency, Austrian Nano Initiative Program, NILaustria project.

REFERENCES

1. I. Utke, P. Hoffmann, and J. Melngailis, J. Vac. Sci. Technol. B26, 1197 (2008).
2. R. Kometani, S. Ishihara, T. Kaito and S. Matsui, Appl. Phys. Express 1, 055001 (2008).
3. E. Platzgummer et al., Proc. SPIE Vol. 7122 (2008) and BACUS Newsletter 25, 2 (Feb 2009).
4. CHARPAN, http://www.charpan.com/.
5. E. Platzgummer et. al, J. Vac. Sci. Technol. B26, 2059 (2008).
6. NILaustria, http://www.nilaustria.at/.
7. J. Fujita, M. Ishida, T. Sakamoto, Y. Ochiai, T. Kaito and S. Matsui, J. Vac. Sci. Technol. B19, 2834 (2001).
8. J.S. Ro, C.V. Thompson, and J. Melngailis, J. Vac. Sci. Technol. B12, 73 (1994).
9. H.-B. Kim, G. Hobler, A. Lugstein, and E. Bertagnolli, J. Micromech. Microeng. 17, 1178 (2007).
10. C. Ebm and G. Hobler, Nucl. Instr. Meth. B (2009) accepted for publication
11. SRIM, http://www.srim.org
12. E. Platzgummer et al., Microelectr. Eng. 83, 936 (2006).
13. G. Hobler, Nucl. Instr. Meth. B96, 155 (1995).

Mater. Res. Soc. Symp. Proc. Vol. 1181 © 2009 Materials Research Society   1181-DD04-01

Swift Heavy Ion Irradiation Induced Effects in Si/SiO$_x$ Multi-Layered Films and Nanostructures

J. W. Gerlach[1], C. Patzig[1], W. Assmann[2], A. Bergmaier[3], Th. Höche[1], J. Zajadacz[1], R. Fechner[1], and B. Rauschenbach[1]

[1]Leibniz-Institut für Oberflächenmodifizierung, Permoserstrasse 15, D-04318 Leipzig, Germany
[2]Ludwig-Maximilians-Universität München, Maier-Leibnitz Laboratory, Am Coulombwall 6, D-85748 Garching, Germany
[3]Universität der Bundeswehr München, Werner-Heisenberg-Weg 39, D-85577 Neubiberg, Germany

ABSTRACT

Amorphous Si/SiO$_x$ multilayered films and nanostructures were deposited on Si substrates by the glancing angle deposition technique using Ar ion beam sputtering of a Si sputter target in an intermittent oxygen atmosphere at room temperature. The chemical composition of the samples was characterized by time-of-flight secondary ion mass spectrometry, as well as - for quantifying these first results - by elastic recoil detection analysis using a 200 MeV Au ion beam. The latter method was found to lead to a significant alteration of the sample morphology, resulting in the formation of complex nanometric structures within the layer stacks. In order to investigate these swift heavy ion irradiation induced effects in more detail, a series of experiments was performed to determine the dominating influences. For this purpose, specific glancing angle deposited multi-layered films and nanostructures were irradiated to constant ion fluence with the same 200 MeV Au ion beam at different incidence angles. Scanning electron microscopy of the stacks before and after swift Au ion irradiation revealed considerable changes in film morphology and density as a function of the ion incidence angle, such as an increased porosity of the silicon layers, accompanied by layer swelling. In contrast, the SiO$_x$ layers did not show such effects, but exhibited clearly visible swift heavy ion tracks. The observed effects became stronger with decreasing ion incidence angle.

INTRODUCTION

When during thin film growth by physical vapor deposition the deposition particle flux impinges the substrate surface under grazing incidence, a new class of non-compact thin films with nano-sized structures emerges. This so-called glancing angle deposition (GLAD) technique, put forward in the last decade by Robbie et al. [1, 2], is based on atomic self-shadowing on the substrate surface. In the beginning of the deposition process, the first nuclei that form on the surface act as seeds for the incoming particle flux. Under low adatom mobility conditions, the substrate region opposite the direction of the incoming atoms is thus shadowed. With ongoing deposition time, this experimental setup enables the fabrication of highly underdense, columnar thin films, with the columns slanted towards the flux direction of the atoms to be deposited. In combination with a controlled substrate rotation during deposition, nanostructures of complex shapes can be "sculpted", resulting in so-called sculptured thin films (STFs). Depending on the ratio of substrate rotation speed to deposition rate, a manifold of different structure shapes, such as helical, zig-zag shaped (chevrons), or vertical columnar structures can be grown with GLAD. A comprehensive review on STFs prepared by GLAD and some applications is given in Ref. [3]. In most cases so far, STFs consisted of only one material, e.g. single elements or compounds.

For the present study, amorphous Si/SiO_x multi-layered films and nanostructures were deposited by GLAD using ion beam sputter deposition (IBSD) in an intermittent oxygen atmosphere at room temperature. With the initial intention to determine the chemical composition of the films, they were characterized amongst others by elastic recoil detection analysis (ERDA), where a 200 MeV $^{197}Au^{15+}$ ion beam was applied. As is already known from many other cases, swift heavy ion irradiation may affect and alter the samples in the irradiated parts (see e.g. ion track formation described in Ref. [4]). The main section of the present contribution focuses on swift heavy ion irradiation induced effects observed with the multi-layered thin films prepared by IBSD. Finally, the obtained results are compared to known models of swift heavy ion irradiation of amorphous and crystalline materials to determine their validity in the present case and in order to explain the observed effects.

EXPERIMENT

Sample preparation

GLAD of Si (for details see Ref. [5]) and Si/SiO_x multi-layered structures and thin films was done by means of Ar ion beam sputter deposition of a Si target under an intermittent O_2 partial pressure. The setup, as depicted in Fig. 1, consists of a high vacuum chamber with a base pressure better than 2×10^{-6} Pa. The chamber is provided with an inductively coupled, radio frequency (13.56 MHz) broad-beam ion source with a focusing triple grid system of 40 mm in diameter and argon as process gas. The ion beam source to target distance measures 15 cm. The substrates are located on a substrate holder attached to an x-y-z manipulator in a distance of 12 cm to the target. The substrate tilt θ (i.e. the deposition angle with respect to the substrate normal) can be adjusted stepless, and the substrate rotation speed (controlled by a computer-driven stepper motor) can be varied from 0.01 rev/min to 0.2 rev/min.

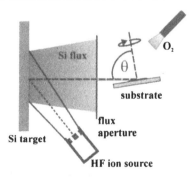

Figure 1. Sample preparation setup.

For this study, the sample preparation was done with an argon flow rate of 4.5 sccm (standard cubic centimeters per minute) through the ion beam source, resulting in a working pressure of approximately 9.0×10^{-3} Pa whilst depositing the Si layers of the multi-layered stacks. The SiO_x layers, on the other hand, were deposited by applying an oxygen flow rate of 20 sccm, thus increasing the total pressure to approximately 2.7×10^{-2} Pa. The Ar ion energy was 1.1 keV, and the experiments were done at room temperature. The deposition of multi-layered STFs with helical shape was done with the deposition angle θ set to 85° and a substrate rotation speed of 0.125 rev/min, whereas Si/SiO_x stacks of dense layers were grown at $\theta = 70°$, and a fast substrate rotation (0.2 rev/min). Examples for both cases can be seen in Fig. 2.

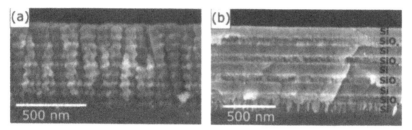

Figure 2. Helical STF (a) and dense multi-layer stack consisting of 5 Si and 4 SiO_x layers (b).

Chemical composition determination

Time-of-flight secondary ion mass spectrometry (TOF-SIMS) was used to obtain the oxygen depth distribution profiles and also to examine the contaminants in the STFs. The samples were analyzed using a pulsed 15 keV Ga^+ ion beam. For depth profiling, an intermittent 2 keV Cs^+ erosion ion beam was applied. Due to the insulating character of the SiO_x layers, charge compensation was inevitable during depth profiling and was accomplished by irradiating the analysis area of the sample with electrons from an electron flood gun. Mass signals representative of the film constituents and contaminations were extracted from the measured negative ion mass spectra. In order to quantify the SIMS depth profiles, elastic recoil detection analysis (ERDA) is a well suited method [6]. For this, the sample is bombarded with a collimated beam of high energetic heavy ions under a small angle of incidence with respect to the sample surface and the energy spectra of forward scattered recoil particles are measured by means of a special ionization chamber with a two-parted anode [7]. ERDA offers the advantages of individual particle identification together with gaining depth information and it benefits from comparable scattering cross sections for light and heavy atoms. The ERDA measurements were performed at the 14 MV tandem accelerator of the Ludwig-Maximilians-University and the Technical University of Munich in Garching, Germany. The ion species used was $^{197}Au^{15+}$ with a kinetic energy of 200 MeV. The angle of incidence with respect to the sample surface was 19°, the detector angle was 37.4°. The pressure in the iso-butane filled ionization chamber was 42 hPa. Calculations of quantitative concentration depth profiles derived from the measured energy spectra were done using the konzERD code [8].

Swift heavy ion irradiation experiments

The samples were irradiated at room temperature with a 200 MeV $^{197}Au^{15+}$ ion beam that was similar to that used for the ERDA measurements. The applied ion fluences were in the range from 3.5×10^{13} to 4.4×10^{13} ions/cm². Ion irradiations at incidence angles of 90°, 75°, 60°, 45°, 30° and 15° with respect to the sample surface were done. Afterwards, the irradiation spots were examined by scanning electron microscopy (SEM) in plan-view and, after cutting the samples through the irradiation spots, in cross-section, using a highly-resolving SEM with a tungsten field-emission electrode and an electron acceleration voltage between 1.5 and 2.5 kV. The working distance was 4 to 5 mm. All micrographs were recorded by using an in-lens secondary electron detector.

RESULTS AND DISCUSSION

Initial observations

A helical Si/SiO$_x$ multi-layered STF (5×Si, 4×SiO$_x$, single layer thickness ~ 60 nm) on a Si substrate was achieved by alternately applying oxygen gas during deposition. First, it was checked by TOF-SIMS, if the built-in oxygen in this heterostructure was distributed as intended. Indeed, the SIMS depth profile in Fig. 3 qualitatively proves the engineered oxygen distribution. The edges within the profiles, where the elemental distribution changes from Si to SiO$_x$ or vice-versa, are remarkably sharp, taking into account the underdense nature of the STF. The dynamic range of the signals corresponding to Si and O is about one order of magnitude. Carbon was found to be the main contaminant in the STF, originating from the residual gas in the chamber, from the graphite extraction grids of the ion beam source, or from the atmosphere after removal of the sample from the vacuum. The inconstant intensity of the C depth profile, that is in unison with the Si depth profile, suggests an alternating C concentration. The ERDA concentration depth profile in Fig. 3 supports the findings and allows quantification hereof. Accordingly, the C concentration in the STF is well below 1 at.% and it decreases with increasing O concentration.

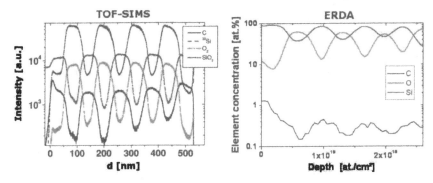

Figure 3. Comparison of TOF-SIMS and ERDA concentration depth profiles of a Si/SiO$_x$ multi-layered STF.

78

Both methods combined prove that the O concentration distribution in the helical multi-layered STF was almost as intended with an x value of ~ 1.8 in the SiO_x layers. The O concentration in the Si layer parts of the multi-layered STF is relatively high with values of more than 10 at.% due to oxide layer formation at the side walls of the helical structures within the STF (see Fig. 4a).

As the optical appearance of the samples changed distinctly within the ion irradiation spots, cross-section SEM was performed in pristine and irradiated parts of the sample (Fig. 4) revealing a severe change of the STF morphology. Before irradiation, the STF consisted of separated helical columns with a significant inter-structure distance, each single helix showing a broadening of its diameter with increasing distance to the substrate. After irradiation (ion fluence of 3.5×10^{13} ions/cm²) however, the former columnar morphology cannot be recognized anymore. While the SiO_x layer parts of the STF show lateral spreading ("glueing") obviously even between neighbored columns once separated in the pristine sample, the Si layers exhibit void formation with a high degree of porosity, accompanied by slight swelling. The effects are so strong that the lowermost SiO_x layer parts of the STF now appear as one completely closed layer. This impression partly diminishes comparing one SiO_x layer with the following one, starting from the lowermost. One striking feature of the porous Si layers is that the voids and the Si structures are elongated fairly along the surface normal and in the upper Si layers rather perpendicularly to ion incidence direction, resulting in a rippled sponge-like arrangement. It should be noted here, that contrary to the Si layers in the multi-layer stack the crystalline Si substrate remained unaltered. According to transmission electron microscopy (TEM) investigations of the pristine and irradiated sample (not shown here) the Si layers, as well as the SiO_x layers were amorphous.

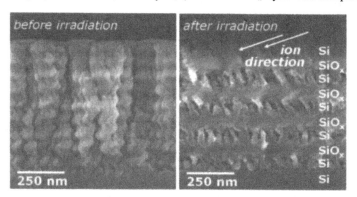

Figure 4. Cross-section SEM micrographs of a helical multi-layered STF before and after 200 MeV Au ion irradiation incident at 19° to the surface.

Swift heavy ion irradiation of compact multi-layer stacks

After the observation of the above described effects, the intricacy of the described helical underdense STF necessitated a simpler model system to investigate the ion irradiation induced effects in more detail. Therefore, a simplified multi-layer stack consisting of compact, dense single layers was prepared (5×Si, 4×SiO$_x$, single layer thickness ~ 75 nm). A series of swift heavy ions irradiations was performed using another ion incidence angle in each irradiation spot while maintaining a constant ion fluence of 4.4×10^{13} ions/cm². The plan-view electron micrographs of the multi-layer stack surface before and after ion irradiation at 15° incidence (Fig. 5) reveal that the observed effects are not restricted to underdense STFs alone, but also appear with compact, dense thin films. While the surface of the pristine sample (the topmost layer in the stack is a Si layer) is relatively smooth, the ripple-like surface in the micrograph of the irradiation spot area appears to be put together of meandering Si straps divided by large voids. In many parts, the projection of the ion incidence direction on the surface tends to be perpendicular to the length axes of these straps.

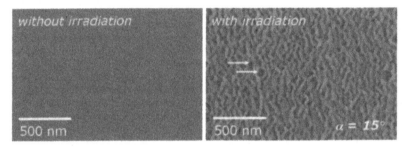

Figure 5. Plan-view SEM micrographs of the surface of a compact multi-layer before and after swift heavy ion irradiation at 15° to the surface.

Cross-section SEM reveals a detailed view of the severe structure changes that fortify with decreasing ion incidence angle (see Fig. 6). For comparison, the SEM micrograph of the pristine film is shown in Fig. 2b. In the case of irradiation at 90° with respect to the sample surface the medium projected range of the Au ions is approximately 21 μm. While the electronic energy loss of the ions within the multi-layer is 17 keV/nm (as obtained by the Monte Carlo simulation code SRIM), the nuclear energy loss of the Au ions, i.e. the energy deposited into the sample by elastic atomic collisions with the sample atoms, is only about 14 eV/nm and therefore negligible. The smaller the incidence angle, the higher is the deposited energy per length along the surface normal, i.e. the higher is the deposited energy density within the single layers in the layer stack. In the SEM micrograph in Fig. 6 corresponding to 90° ion incidence angle there is no sign of a change in the Si layers. A closer look at the SiO$_x$ layers reveals lines within the layers that are parallel to the ion incidence direction. These lines may be related to ion tracks, that form in the SiO$_x$ layers along the ion trajectories, and are also visible in the sample parts irradiated at other angles. According to Ref. [4], for 17 keV/nm electronic energy loss in SiO$_2$, the effective radius of an amorphous ion track is about 4 nm. The micrograph of the 75° irradiation case is very similar to the one of the 90° irradiation. Void formation in the Si layers is first observed at 60° incidence angle where the voids appear almost circular in the micrograph. From 45° to 15°, the voids grow and change in shape, enlarging mostly in the direction perpendicular to the ion incidence direction. Simultaneous to the void formation, the Si layers swell due to conservation of the Si volume. In the case of 15° ion incidence, the average Si layer thickness is about 170 to 180 % of the initial thickness before irradiation, whereas the SiO$_x$ layers do not exhibit any swelling at all. For the 15° case the corresponding electronic energy loss of the Au ions projected on the sample normal increased to about 65 keV/nm, i.e. the energy density within the single layers of the stack is almost four times higher than in the case of perpendicular irradiation. The corresponding medium projected ion range of 5.5 µm is still much larger than the total thickness of the swollen multi-layer stack (approximately 900 nm).

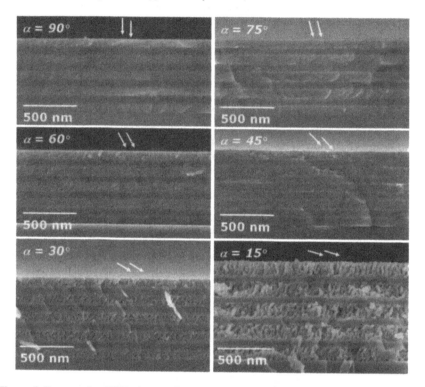

Figure 6. Cross-section SEM micrographs of a compact Si/SiO$_x$ multi-layer irradiated at different ion incidence angles (see white arrows) with a constant fluence of 4.4×10^{13} ions/cm².

Swift heavy ion irradiation of SiO$_x$ nano walls

Although obviously the more pronounced ion irradiation induced effects took place within the Si layers, a further step was done to learn more about how the ion irradiation affects the SiO$_x$ layers. For this purpose, a compact 500 nm thick SiO$_x$ film was deposited on a Si substrate. Then, the film was structured by electron beam lithography (EBL) and subsequent reactive low-energy Ar ion beam sputtering. In the end, parallel SiO$_x$ nano walls (100, 300 and 500 nm wide) remained in the EBL-structured 100 × 100 µm² large areas of the SiO$_x$ film. These nano wall fields were irradiated with the 200 MeV Au ion beam at an angle of 15° to the sample surface, with the projection of the ion beam incidence direction on the sample surface lying perpendicular to the nano walls. Thus, the ion incidence angle was 15° with respect to the tops of the nano walls and 75° with respect to the sides of the nano walls. The ion fluence was again 4.4×10^{13} ions/cm². The plan-view SEM micrograph in Fig. 7 shows the tops of (in this case about 500 nm wide) nano walls without and with ion irradiation. While the pristine SiO_x nano walls (left) appear to be relatively perpendicular to the substrate surface, the irradiated nano walls (right) show another contrast suggesting that the walls here are not perpendicular to the surface anymore, but rather tilted away from the ion incidence direction.

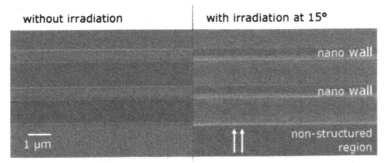

Figure 7. Plan-view SEM micrographs of pristine (left) and swift heavy ion irradiated (right) SiO_x nano walls. The arrows indicate the projected ion beam incidence direction.

That this tilt of the walls is indeed the case can be clearly seen in Figs. 8 and 9. Viewed at an oblique angle with respect to the sample surface, a cut through a nano wall field (cut perpendicularly to the long side of the nano walls) is shown in Fig. 8. With the Au ion beam incidence having been from the left side in this micrograph, the nano walls are found to be bent towards the right side of the micrograph, i.e. away from the ion incidence direction.

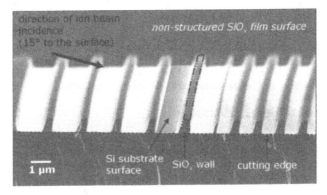

Figure 8. SEM micrograph at an oblique angle of view showing the cut through a swift heavy ion irradiated SiO_x nano wall field.

The right part of Fig. 8 was recorded with higher magnification to clearly demonstrate the bent nano walls. In a rough approximation, the averaged tilt angle obtained from cross-section micrographs of the walls lies between 15° and 20° with respect to the sample normal.

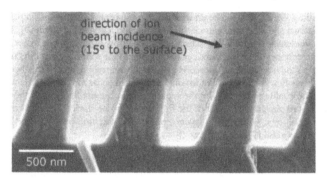

Figure 9. Higher magnification micrograph of the right part of the SEM micrograph in Fig. 8.

The right part of Fig. 8 was recorded with higher magnification to clearly demonstrate the bent nano walls. In a rough approximation, the averaged tilt angle obtained from cross-section micrographs of the walls lies between 15° and 20° with respect to the sample normal. These results give a hint, how the SiO_x layers inside an ion irradiated Si/SiO_x multi-layer stack react. It can be deduced that the Si layers in between the deformed SiO_x layers will be affected by this deformation, too.

The main observations made in the present study can be summarized as follows:
a) Swift heavy ion irradiation of amorphous Si and SiO_x layers in multilayered STFs resulted in void formation in the Si layers and lateral spreading within the SiO_x layers.
b) Swift heavy ion irradiation of amorphous Si and SiO_x layers in compact, dense multi-layered thin film stacks lead to void formation in the Si layers, accompanied by Si layer swelling normal to the sample. Lateral spreading of the SiO_x layers was not observed, as the layers were compact and closed already before the ion irradiation; the SiO_x layers exhibited no swelling normal to the sample.
c) Swift heavy ion irradiated thin SiO_x nano walls tilted away from the ion incidence direction.
d) The single crystalline Si substrates, on which the multi-layered STFs or compact films were deposited, did not show any ion irradiation induced change.

Further observations not explicitly reported in the present paper are:

e) The Si layers in Si/SiO_x and SiO_x/Si double layers with thicknesses of 200 nm/500 nm and 500 nm/300 nm, respectively, did not exhibit void formation.

Although the amount of data is still insufficient for a final explanation of the observed effects, a short, rather premature discussion shall be given at this point, taking into account results reported in the literature. It seems clear that the observed effects are a result of a complex combination of more or less well understood effects, that all have their origin in the irradiation of matter with highly energetic ions.

Concerning swift heavy ion beam irradiation of amorphous materials, in particular one effect is well known and well understood: the so-called ion hammering effect [9]. This effect comprises ion irradiation induced, non-saturating plastic flow of amorphous materials in such a way that the amorphous material expands perpendicular to the ion incidence direction and shrinks parallel to the ion incidence direction [10]. If the amorphous material is bound to a substrate, this leads to a swift heavy ion irradiation induced shear movement [9]. While this effect seems to be qualitatively valid for the separated straps in the Si layers of the multilayer stacks, it does not explain the behavior of the SiO_x layers in the stacks. The latter may be caused by the under-stoichiometry of the SiO_x layers, because for stoichiometric SiO_2 the ion hammering effect is valid [9].

Formation of porous Si by swift heavy ion irradiation was recently observed by Hedler et al. [11]. They not only identified the origin of the deformation of the layers to be the ion hammering effect [12], but they also observed a dramatic density change within the pre-amorphized Si layers. Astonishingly, although irradiating with a 600 MeV Au ion beam at an electronic energy loss of 21 keV/nm, the ion fluence required to obtain the effect was higher by at least one order of magnitude than the fluence used in the present study. This may be an enhancement effect that is provoked by separating the amorphous Si by amorphous SiO_x layers.

Void formation and increasing porosity due to swift heavy ion irradiation was also observed in Ge already a decade earlier by Huber et al. [13]. Interestingly, in a recent report on 700 keV Kr ion implanted amorphous Ge, where the nuclear energy loss is dominant, void formation and swelling was observed, too [14]. The authors attributed these effects to compressive stress induced, interstitial-mediated viscous flow in combination with well-localized vacancy defects and successfully reproduced their experiment by molecular dynamics simulations. Obviously, the tendency for void formation and swelling seems to be a universal property of amorphous Ge and Si.

Back to the present study, an effect that can be ruled out, is the formation of voids in Si due to gas bubbles forming in the layers. If this would be the case, then in the swift heavy ion irradiated Si/SiO_x and SiO_x/Si double layers reported of in the results summary above there should have emerged voids, resulting in a sponge-like Si layer. But this was not the case.

One side effect whose impact can only be speculated about is swift heavy ion induced mixing of the Si/SiO_x layer stacks. It is well known that such effects can appear, when bi-layers or multi-layer stacks are irradiated with swift heavy ions [15]. The Si and SiO_x layers in the present stacks were relatively thin compared to the averaged length of ion trajectories, implying an intermixing of the layers.

The fact that the SiO_x layers were not stoichiometric may have lead to further deviations from stoichiometry by swift heavy ion induced material transport, e.g. it seems possible that the SiO_x layers tend to achieve SiO_2 stoichiometry by release of Si atoms. A rather similar effect was observed by Carlotti et al. who reported the formation of Si bumps at the interface of swift heavy ion irradiated silicon oxide films on Si [16], while Arnoldbik et al. found O_2 desorbing from swift heavy ion irradiated SiO_2 films [17].

Besides others, a large influence of the swift heavy ion irradiation geometry on the dimensional changes of SiO_2 was reported by Benyagoub et al. [18]. If a deviation from stoichiometry will enhance or diminish this trend, cannot be answered so far, as there is not enough experimental data available for non-stoichiometric SiO_x samples. A high correlation between the angle of ion incidence and plastic deformation of the irradiated amorphous solids was also found in Ref. [19].

Another type of effect is the expectable formation of mechanical stress at the interfaces of the single layers due to the strong swift heavy ion driven deformations of the Si and SiO_x layers. This stress may in turn enhance others of the above mentioned effects. As an example, Trautmann et al. could influence track formation by applying external mechanical stress to the irradiated samples [20].

Finally, the influence of the substrate is not to be neglected, as the results indicate an increase of the observed effects like swelling of Si or lateral expansion of SiO_x in layers nearer to the substrate.

CONCLUSIONS

Driven by swift heavy ion irradiation, severe morphological changes in under-dense Si/SiO_x multi-layered sculptured thin films and compact thin film stacks were achieved. The amorphous Si and SiO_x single layers in the multi-layer stacks were affected completely different, leading to sponge-like Si layers between laterally expanded SiO_x layers. With lower ion incidence angle with respect to the samples the effects were observed to become larger. The ion fluence required to obtain these effects was relatively low in comparison to values reported in the literature. A detailed fluence series will be necessary to gain more detailed information on the mechanism of void formation in the Si layers and to figure out if there exists something like an incubation fluence that must be surpassed for the effects to take place. Further, experiments concerning the irradiation effect on sole thin Si and SiO_x layers, as well as on thin Si/SiO_x or SiO_x/Si double layers are required to investigate the influence of single layer thickness and interface stress on the swift heavy ion irradiation induced effects.

ACKNOWLEDGMENTS

The authors would like to thank S. Klaumünzer (Helmholtz-Zentrum Berlin für Materialien und Energie, Germany) for valuable discussions and for imparting a deeper insight into the mechanisms of the ion hammering effect, W. Wesch (University of Jena, Germany) for providing input concerning swift heavy ion induced effects in amorphous Ge, as well as M. Toulemonde (CIMAP, Laboratoire CEA-CNRS-ENSICAEN-Université de Caen, Caen, France) for bringing in his vast expertise in the field of swift heavy ion induced effects in matter.

REFERENCES

1. K. Robbie, M.J. Brett, and A. Lakhtakia, *J. Vac. Sci. Technol.* **A13**, 2991 (1995).
2. A. Lakhtakia, R. Messier, M.J. Brett, and K. Robbie, *Innov. Mater. Res.* **1**, 165 (1996).

3. M.M. Hawkeye and M.J. Brett, *J. Vac. Sci. Technol.* **A25**, 1317 (2007).
4. M. Toulemonde, S. Bouffard, and F. Studer, *Nucl. Instrum. Meth.* **B91**, 108 (1994).
5. C. Patzig, B. Rauschenbach, W. Erfurth, and A. Milenin, *J. Vac. Sci. Technol.* **B25**, 833 (2007).
6. W. Assmann, H. Huber, Ch. Steinhausen, M. Dobler, H. Glückler, and A. Weidinger, *Nucl. Instrum. Meth.* **B89**, 131 (1994).
7. W. Assmann, *Nucl. Instrum. Meth.* **B64**, 267 (1992).
8. A. Bergmaier, G. Dollinger, and C.M. Frey, *Nucl. Instrum. Meth.* **B99**, 488 (1995).
9. S. Klaumünzer, *Nucl. Instrum. Meth.* **B215**, 345 (2004).
10. S. Klaumünzer, Changlin Li, S. Löffler, M. Rammensee, G. Schumacher, and H. Ch. Neitzert, *Radiat. Eff. Defect. S.* **108**, 131 (1989).
11. A. Hedler, S. Klaumünzer, and W. Wesch, *Nucl. Instrum. Meth.* **B242**, 85 (2006).
12. A. Hedler, S. Klaumünzer, and W. Wesch, *Phys. Rev.* **B72**, 054108 (2005).
13. H. Huber, W. Assmann, R. Grötzschel, H.D. Mieskes, A. Mücklich, H. Nolte, W. Prusseit, *Mater. Sci. Forum* **248-249**, 301 (1997).
14. S.G. Mayr and R.S. Averback, *Phys. Rev.* **B71**, 134102 (2005).
15. W. Bolse, *Mater. Sci. Eng.* **R12**, 53 (1994).
16. J.-F. Carlotti, A.D. Touboul, M. Ramonda, M. Caussanel, C. Guasch, J. Bonnet, and J. Gaslot, *Appl. Phys. Lett.* **88**, 041906 (2006).
17. W.M. Arnoldbik, N. Tomozeiu, and F.H.P.M. Habraken, *Nucl. Instrum. Meth.* **B219**, 312 (2004).
18. A. Benyagoub, S. Löffler, M. Rammensee, S. Klaumünzer, G. Saemann-Ischenko, *Nucl. Instrum. Meth.* **B65**, 228 (1992).
19. A. Audouard, J. Dural, M. Toulemonde, A. Lovas, G. Szenes, and L. Thomé, *Phys. Rev.* **B54**, 15690 (1996).
20. C. Trautmann, S. Klaumünzer, and H. Trinkaus, *Phys. Rev. Lett.* **17**, 3648 (2000).

Mater. Res. Soc. Symp. Proc. Vol. 1181 © 2009 Materials Research Society   1181-DD04-02

My Modeling Nanocluster Formation During Ion Beam Synthesis

Chun-Wei Yuan, [1,2] Diana O. Yi, [1,2] Ian D. Sharp, [3] Swanee J. Shin, [1,2] Christopher Y. Liao, [1,2] Julian Guzman, [1,2] Joel W. Ager III, [2] Eugene E. Haller[1,2] and Daryl C. Chrzan[1,2]

[1]Department of Materials Science and Engineering, University of California, Berkeley, CA 94720, U.S.A.
[2]Materials Science Division, Lawrence Berkeley National Laboratory, Berkeley, CA 94720-1760, U.S.A.
[3]Walter Schottky Institut, Technische Universität München, Am Coulombwall 3, 85748 Garching, Germany

ABSTRACT

Ion beam synthesis of nanoclusters is studied via both kinetic Monte Carlo simulations and the self-consistent mean-field solution to a set of coupled rate equations. Both approaches predict a steady-state shape for the cluster size distribution that depends only on a characteristic length determined by the ratio of the effective diffusion coefficient times the effective solubility to the ion flux. The average cluster size in the steady state regime is determined by the implanted species/matrix interface energy.

INTRODUCTION

Ion beam synthesis (IBS) is a technologically important method to synthesize nanocrystals within a solid matrix. The technique involves embedding one or more strongly-segregating species into a suitable matrix through the implantation of energetic ions. During implantation, the ions move about and encounter other implanted ions, leading to cluster nucleation. The clusters may then grow/evaporate via adsorption/desorption of single atoms. The description thus far resembles a 3-D analogue of 2-D submonolayer epitaxial nucleation, growth and coarsening [1-8].

IBS, however, differs from classical nucleation and growth problems in one important respect: During IBS, cluster growth is often interrupted by ion damage. The deposition of ions fragments the clusters, leads to inverse Ostwald ripening [9], and ultimately limits cluster sizes. The basic understanding of IBS has been exploited to fabricate interesting nanostructures and alter cluster size distributions [9-16]. But unlike submonolayer epitaxy [2-8] and low dose semiconductor doping via implantation that are extensively studied and understood [17-19], no detailed, comprehensive, quantitative theory of IBS has been developed.

In this study, kinetic Monte Carlo (KMC) simulations, and the mean-field self-consistent solution to a set of coupled rate equations are developed to model cluster size distribution evolution during implantation. It is shown that the two key parameters governing the as-implanted size distribution for a given material are the characteristic length, $L = (Dn_\infty / F)^{1/2}$, where F is the volumetric flux and n_∞ is the ion's solubility, and γ the interface energy between the implanted species and the matrix.

THEORY

The employed KMC simulations include five fundamental processes: 1) implantation into an amorphous silica matrix at a rate F, 2) the off-lattice random walk of implanted atoms, represented by a diffusion coefficient D (the hop distance is chosen to be 5 Å), 3) the attachment of atoms to each other and existing clusters followed by immediate relaxation of all clusters to a spherical shape, 4) the thermally driven detachment of atoms from existing clusters, and 5) the ion damage induced fragmentation of clusters.

The binding of atoms to clusters (and each other) is modeled as described in reference [21]. The radius of an atom is chosen to be $r_a = (3\Omega/4\pi)^{1/3}$, with Ω the equilibrium atomic volume of the implanted species. An atom binds with another implanted atom or existing cluster when the two "touch." In order to accelerate the simulations, the atomic scale structure of the clusters is not stored. Instead, all clusters are assumed to relax immediately to a spherical shape with the volume equal to the total volume of the constituent atoms.

Thermodynamic desorption from existing clusters is described by a local equilibrium approximation [3]. The desorption rate depends on the interface energy between the matrix and the implanted species, γ. Specifically, desorption from a cluster containing s-atoms is presumed to occur at a rate $\dfrac{1}{\tau_s} \propto \sigma_s n_\infty \exp\dfrac{2\gamma\Omega}{kTR_s}$, with R_s the radius of a cluster containing s atoms, σ_s a capture length for clusters containing s atoms, n_∞ the solubility, and kT Boltzmann's constant multiplied by temperature. Clusters are represented as spheres at all times. When atoms desorb/adsorb from/to existing clusters, the center of mass of the cluster and desorbed atom remains fixed. Ion damage is modeled by extending the approach of Heinig and collaborators [9]. When a high-energy ion impacts a solid, it leads to a collision cascade. The net effect of this cascade is to displace a large number of atoms in random directions. Heinig et al. suggest that this cascade be modeled by displacing the atoms within a cluster in random directions with distances r governed by a Poisson distribution: $p(r) = \dfrac{1}{8\pi\lambda^3}\exp(-r/\lambda)$. The value of λ is then determined by fitting the displacement profiles for an ion-damaged embedded slab computed using TRIM [21]. Typical values of λ range between 3 and 5 Å.

An impinging ion hits existing nanoclusters with a probability proportional to the volume of the nanocluster, and the ion flux. When a cluster is hit, every atom in the cluster is displaced according to the aforementioned Poisson distribution. The atoms are immediately checked to see if they "touch" other atoms. All "touching" atoms bind into or to clusters, and the simulation proceeds. This coarse-grained description of the cascade is computationally efficient and provides a good description of observations. Specifically, molecular dynamics simulations of Au ions impacting free standing Au clusters yield a fragment size distribution that is characterized by a power law distribution of small fragments, $f_s \sim s^{-\alpha}$, with f_s the number of fragments with s atoms, and $\alpha = 2.3$ [22]. Modeling the same process using KMC renders the same exponent.

A set of coupled rate equations (RE) are derived following references [2, 3, 20] while adding the effects of cluster fragmentation. Scaling all lengths by L and all times by n_∞/F, the relevant equations become:

$$\frac{d\langle \tilde{n}_1 \rangle}{d\tilde{t}} = \tilde{F} - 2\tilde{\sigma}_1 \langle \tilde{n}_1 \rangle^2 - \sum_{j>1} \tilde{\sigma}_j \langle \tilde{n}_j \rangle \langle \tilde{n}_1 \rangle + 2\frac{\langle \tilde{n}_2 \rangle}{\tilde{\tau}_2} + \sum_{j>2} \frac{\langle \tilde{n}_j \rangle}{\tilde{\tau}_j} + \tilde{F}\tilde{\Omega}\sum_{j>1} \langle \tilde{n}_j \rangle (j+1)K_1(\alpha, j) \qquad (1)$$

and

88

$$\frac{d\langle\tilde{n}_s\rangle}{d\tilde{t}} = \tilde{\sigma}_{s-1}\langle\tilde{n}_{s-1}\rangle\langle\tilde{n}_1\rangle - \tilde{\sigma}_s\langle\tilde{n}_s\rangle\langle\tilde{n}_1\rangle - \frac{\langle\tilde{n}_s\rangle}{\tilde{\tau}_s} + \frac{\langle\tilde{n}_{s+1}\rangle}{\tilde{\tau}_{s+1}} - \tilde{F}\tilde{\Omega}\langle\tilde{n}_s\rangle(s+1) + \tilde{F}\tilde{\Omega}\sum_{j>s}\langle\tilde{n}_j\rangle(j+1)K_s(\alpha,j).\quad (2)$$

In Eqs. (1) and (2), $\langle\tilde{n}_s\rangle$ is the dimensionless average density of clusters containing s atoms, $\tilde{\sigma}_s$ is the dimensionless capture length for an s-atom cluster to capture a diffusing atom. The dimensionless thermally activated desorption time for an atom to leave a cluster of s atoms is $\tilde{\tau}_s$. The last terms of Eqs. (1) and (2) represent the fragmentation process: $\tilde{\Omega}$ is the dimensionless atomic volume of the implanted species and the term $\langle\tilde{n}_j\rangle\tilde{\Omega}(j+1)$ gives the probability of collision between an implanted atom and a j-cluster. $K_s(\alpha,j)$ is the number of s-clusters generated as a result of a fragmented j-cluster: $K_s(\alpha,j)=\left(\frac{j}{H_{j-1}^{1-\alpha}}\right)s^{-\alpha}$, with $\alpha=2.3$, and H_j^r the jth harmonic number of order r.

DISCUSSION AND RESULTS

For an initial calculation, we consider parameters near those appropriate for the implantation of Ge into amorphous silica. The initial volumetric implantation rate is set to be $F = 10^{-6}$ sec^{-1} Å^{-3}, roughly a factor of 7 larger than that employed by Sharp et al. [23], a choice made to enable reasonable computation times. TRIM calculations for 120 keV Ge implanted into SiO2 give λ = 3.5 Å. The diffusion coefficient for Ge in SiO$_2$ at room temperature is taken as $D = 6.5 \times 10^{-10}$ cm^2 s^{-1} while the solid solubility is $n_\infty = 10^8$ cm^{-3}, both approximately equilibrium values [24]. Experimental studies of stress relaxation [11] indicate that the Ge/SiO$_2$ interface energy is near 0.7-0.9 J m^{-2}. However, choosing interface energies this large increases the critical cluster size for nucleation, and with it the required computation resources. Thus the Ge/silica interface energy is set to 0.2 J m^{-2}.

Figure 1 shows a typical comparison between KMC and the rate equation results for the evolution of a cluster size distribution during implantation. The two computational approaches display qualitative agreement during the first two stages of growth. During the very early stage of growth the effect of damage cascade is negligible and the cluster size distribution attains a characteristic swept forward shape. As more mass is introduced into the system, fragmentation events occur more frequently and a peak representing the small fragments appears. This peak continues to grow and eventually dominates, rendering a unimodal distribution weighted at smaller sizes. It is observed empirically that the cluster size distribution reaches a steady-state. As this limit is approached, the shape of the size distribution varies slowly and the steady-state shape serves as a good description of the shape of the cluster size profile during the latter stages of IBS.

In s-space, the scaled steady-state cluster size distribution, $g\left(\frac{s}{\bar{s}}\right)$ [20], can be found by

solving: $0 = \frac{d}{dt}g\left(\frac{s}{\bar{s}}\right) = \frac{d}{dt}\left(\bar{s}\frac{s\langle n_s\rangle}{Ft}\right)$, with $\bar{s} = \frac{1}{Ft}\sum_{s=2}^{\infty}s^2\langle n_s\rangle$, in the limit that $\frac{d\bar{s}}{dt} \approx 0$.

Combining this condition with Eqs. (1) and (2) and solving for large times yields the steady-state cluster size distribution. After conducting intermediate asymptotic analyses according to [25], one concludes that the steady-state solution depends on two parameters: the dimensionless atomic volume, $\tilde{\Omega}$, and the interface energy, γ. In practice, however, we find that the steady-state shape of the distribution depends only on $\tilde{\Omega}$ and the average size depends on γ. This behavior contrasts that observed during 2-D epitaxial growth in which the distribution shape is determined by the critical cluster size, a quantity that depends sensitively on γ.

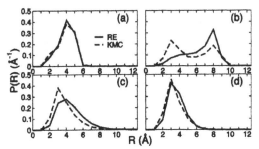

Figure 1. KMC (dashed line) and RE (solid line) size distributions at (a) 3 seconds, (b) 30 seconds, (c) 300 seconds and (d) 3000 seconds for the same parameters as the plots shown in Figure 1. The data from the rate equations has been binned using the same binwidth as used for KMC simulations in order to facilitate comparison.

Experimentally measured as-implanted size distributions can be fitted to theory to obtain an estimate for the value of L. Figure 2 displays the fitted curves for three materials. Ge is best fit with a value of $L = 1.38$ Å, Co with $L = 0.08$ Å, and Ag with $L = 2.66$ Å [28]. For the case of Ge, assuming that coarsening is very slow (as is observed experimentally [10]), we can compare the observed value with the expected value. Using the accepted (normal) diffusion coefficient for Ge in silica one predicts $L = 0.052$ Å. The measured value for L exceeds the initial prediction by a factor of ≈ 26.4. The implication is that the ratio Dn_{∞}/F exceeds the expected value by a factor of ≈ 700. This increase is most likely due to ion damage within the matrix and is within the range expected for transient enhanced diffusion (TED) [19]. Hence, the values of L needed to model experiments are reasonable. Further, measurements of the shape of the cluster size distribution can be used to determine the value of the transient enhanced diffusion coefficient during implantation. It is unclear at this point how the value of L varies with dose during implantation. The value of solid solubility in SiO_2 is difficult to assess at temperatures relevant to IBS [24]. While little research has been done on TED in SiO_2, studies of B implanted into Si suggest saturation of TED at doses greater than $\approx 3 \times 10^{14}$ cm^{-2} [26], roughly 1% of typical IBS doses. We expect, therefore, that for the systems studied here, L changes rapidly during the initial stages of growth, but quickly stabilizes to the value deduced from the steady-state distributions. Further, ion damage serves to minimize the influence of the initial implantation performed during the stages with rapidly varying L.

The theory also suggests how one might narrow IBS cluster-size distributions. Figure 3(b) plots the computed full-width at half-maximum, scaled by the average cluster radius, $\Delta R / \langle R \rangle$, as a function of L for Ge. As L decreases, so does the relative width. Near $L \approx 1$ Å the curve shows a sharp downward trend. Experimentally, L can be decreased by increasing the flux of ions and/or decreasing the transient enhanced diffusion, perhaps by cooling [19, 27].

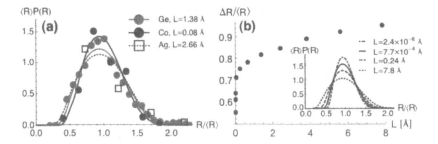

Figure 2. (a) Steady-state shape of the nanocrystal size distribution function obtained by fitting theory to experimental room-temperature as-implanted size distributions for Ge, Co [12] and Ag [13]. The fits provide an excellent description of the available data. (b) Full-width at half maximum plotted as a function of L predicted by the model. The inset shows the steady-sate cluster size distributions expected for different L.

CONCLUSIONS

In conclusion, a theory for the shape of the cluster-size distributions arising during IBS is introduced and compared with the results of kinetic Monte Carlo simulations. The theory predicts that the nanocluster size distribution asymptotes to a steady-state profile, the shape of which depends only on $\tilde{\Omega}$, whereas the average size cluster is determined by the interface energy. The theory can be used to measure the value of transient enhanced diffusion coefficient times effective solubility during implantation, and to develop processing routes to narrow cluster-size distributions.

ACKNOWLEDGMENTS

This research is supported by the Directorate, office of Science, Office of Basic Energy Sciences of the U.S. Department of Energy under Contract No. DE-AC02-05CH11231.

REFERENCES

1. L. Ratke and P. W. Voorhees, *Growth and Coarsening*, 1st ed. (Springer-Verlag, New York, 2002).
2. G. S. Bales and D. C. Chrzan, *Phys. Rev. B* **50**, 6057 (1994).
3. G. S. Bales and A. Zangwill, *Phys. Rev. B* **55**, 1973 (1997).
4. J. A. Stroscio and D. T. Pierce, *Phys. Rev. B* **49**, 8522 (1994).
5. J. G. Amar and F. Family, *Phys. Rev. Lett.* **74**, 2066 (1995).
6. C. Ratsch, A. Zangwill, P. Smilauer, and D. D. VVedensky, *Phys. Rev. Lett.* **72**, 3194 (1994).
7. Y. W. Mo, J. Kleiner, M. B. Webb, and M. G. Lagally, *Phys. Rev. Lett.* **66**, 1998 (1991).
8. A. Pimpinelli, J. Villain, and D. E. Wolf, *Phys. Rev. Lett.* **69**, 985 (1992).
9. K. H. Heinig, T. Müller, B. Schmidt, M. Strobel, and W. Möller, *Appl. Phys. A* **77**, 17 (2003).

10. Q. Xu, I. D. Sharp, C. W. Yuan, D. O. Yi, A. M. Glaeser, C. Y. Liao, A. M. Minor, J. W. Beeman, M. C. Ridgway, J. W. Ager III, D. C. Chrzan, et al., *Phys. Rev. Lett.* **97**, 155701 (2006).
11. I. D. Sharp, D. O. Yi, Q. Xu, C. Y. Liao, J. W. Beeman, Z. Liliental-Weber, K. M. Yu, D. Zhakarov, J. W. Ager III, D. C. Chrzan, et al., *Appl. Phys. Lett.* **86**, 063107 (2005).
12. E. Cattaruzza, F. Gonella, G. Mattei, P. Mazzoldi, D. Gatteschi, C. Sangregorio, M. Falconieri, G. Salvetti, and G. Battaglin, *Appl. Phys. Lett.* **73**, 1176 (1998).
13. C. Z. Jiang and X. J. Fan, *Surface and Coatings Technology* **131**, 330 (2000).
14. G. D. Marchi, G. Mattei, P. Mazzoldi, and C. Sada, *J. Appl. Phys.* **92**, 4249 (2002).
15. R. Giulian, P. Kluth, L. L. Araujo, D. J. Llewellyn, and M. C. Ridgway, *Appl. Phys. Lett.* **91**, 093115 (2007).
16. V. Ramaswamy, T. E. Haynes, C. W. White, W. J. MoberlyChan, S. Roorda, and M. J. Aziz, *Nanoletters* **5**, 373 (2005).
17. G. Dearnaley, *Nature* **256**, 701 (1975).
18. T. Shinada, S. Okamoto, T. Kobayashi, and I. Ohdomari, *Nature* **437**, 1128 (2005).
19. N. E. B. Cowern and C. S. Rafferty, *MRS Bulletin* **25**, 39 (2000).
20. D. O. Yi, M. H. Jhon, I. D. Sharp, Q. Xu, C. W. Yuan, C. Y. Liao, J. W. Ager III, E. E. Haller and D. C. Chrzan, *Phys. Rev. B* **78**, 245415 (2008).
21. J. F. Ziegler, J. P. Biersack, and U. Littmark, *The Stopping and Range of Ions in Solids* (Pergamon Press, New York, 1985).
22. R. Kissel and H. M. Urbassek, *Nucl. Inst. Meth. Phys. Res. B* **180**, 293 (2001).
23. I. D. Sharp, Q. Xu, C. Y. Liao, D. O. Yi, J. W. Beeman, Z. Lilienthal-Weber, K. M. Yu, D. N. Zakharov, J. W. Ager III., D. C. Chrzan, and E. E. Haller, *J. Appl. Phys.* **97**, 124316 (2005).
24. J. D. McBrayer, R. M. Swanson, and T. W. Sigmon, *J. Electrochem. Soc.* **133**, 1242 (1986).
25. G. I. Barenblatt, *Scaling*, 1st ed. (Cambridge University Press, 2003).
26. N. E. B. Cowern, K. T. F. Janssen, and H. F. F. Jos, *J. Appl. Phys.* **68**, 6191 (1990).
27 L. Pelaz, G. H. Gilmer, V. C. Venezia, H.-J. Gossman, M. Jaraiz, and J. Barbolla, *Appl. Phys. Lett.* **74**, 2017 (1999).
28. Due to the small number of points describing this distribution, the uncertainty in the value of L for Ag is substantial.

Mater. Res. Soc. Symp. Proc. Vol. 1181 © 2009 Materials Research Society   1181-DD04-03

James R. Groves[1,3], Robert H. Hammond[2], Ann F. Marshall[2], Raymond F. Depaula[3], Liliana Stan[3] and Bruce M. Clemens[1]

[1]Department of Materials Science and Engineering, Stanford University, Stanford CA 94305
[2]Geballe Laboratory for Advanced Materials, Stanford University, Stanford CA 94305
[3]Superconductivity Technology Center, Los Alamos National Laboratory, Los Alamos, NM 87545

ABSTRACT

The use of an ion beam assist during the concurrent deposition of cubic materials can result in the growth of crystallographically oriented thin films. A model system, magnesium oxide (MgO), has been successfully used as a biaxially textured template film and develops texture in a different manner from that of other well-studied materials, like yttria-stablized zirconia. Here, we present data on the initial nucleation of biaxial texture in this model system using a novel in-situ quartz crystal microbalance (QCM) substrate combined with in-situ reflected high-energy electron diffraction (RHEED). Temporal correlation of mass uptake with the RHEED images of the growing surface can be used to elucidate the mechanism of texture development in these films. Experimental data shows that the initially polycrystalline MgO film develops biaxial crystallographic texture at a thickness of ~2 nm, regardless of the ion-to-molecule ratio. RHEED images show the onset of texture occurs quickly and is somewhat analogous to a solid phase re-crystallization process with crystallite sizes of ~3 to 4 nm. Imaging with transmission electron microscopy has corroborated these observations. Changes in the ion-to-molecule ratio can influence the crystallite size and affect the nucleation density of these films. Growth of these films on various substrates changes the sticking coefficient of the MgO and influences the nucleation density and film growth mode as well. This opens the possibility of using MgO and other materials to develop biaxially textured crystallites with a narrow, specified size distribution for nanoscale applications.

INTRODUCTION

Traditional ion beam assisted deposition (IBAD) is used to modify the density, grain size, microstructure and surface roughness of polycrystalline thin films for applications like tribological and corrosion protection coatings[1-3]. The deposition of Nb with concurrent off-normal incident ion beam irradiation was one of the first reports to show that both the in-plane and out-of-plane orientation of a growing film could be influenced using IBAD[4]. Iijima and co-workers in Japan found that they could deposit biaxially oriented thin films of yttria-stabilized zirconia (YSZ), a well-known radiation damage tolerant material, using low energy IBAD processing and then use the resultant film as a template layer for the subsequent heteroepitaxial deposition of the high temperature superconductor (HTS) $YBa_2Cu_3O_7$ (YBCO)[5]. Further advancements were made in IBAD technology with the introduction of IBAD MgO[6]. Researchers at Stanford University showed that similar quality biaxially textured MgO films, only 10 nm thick, have comparable crystallographic texture to 1mm thick YSZ[7]. Using MgO as a template for subsequent YBCO deposition enabled the production of long lengths of the superconductor on metal substrates for power distribution applications[8]. The majority of advancement in the field of biaxially textured template layers has been focused on improving the IBAD process for the industrial manufacture HTS wires.

Scientifically, the IBAD process presents the ability to manipulate a growing film on the nanometer scale. Indeed, many researchers in the field have developed theories to explain the development of the texture in these two well-characterized oxides: YSZ and MgO[9, 10]. However, there is a distinct difference between the development of texture in the two materials as evidenced by the disparity between the amounts of each material required to achieve useful in-plane alignment. In the case of YSZ, investigators have shown that the process is one of competitive grain growth dynamics: so-called grain evolution[11]. Several theories have been proposed to explain the mechanism of IBAD MgO texture development. Wang et al. proposed that oriented grains survive due to differential ion beam sputtering and, upon coalescence; these grains undergo a minimization of grain boundary energy that results in grain-to-grain alignment through relative grain rotations [12]. A second similar model, proposed by Usov and co-workers, depends upon damage anisotropy between crystallographic planes in the grains where orientations with higher tolerances for damage accumulation survive and grow at the expense of other grains resulting in a preferred orientation[13]. Brewer and Atwater theorize that oriented grains grow through induced solid phase crystallization by ion beam bombardment of an initially amorphous MgO layer [14]. Whatever the mechanism, it appears that the irradiation of MgO with a low energy ion beam directed along a specific crystallographic direction can produce biaxial texture at or very near the nucleation of the film on the surface.

In a recent paper, we used a specially built quartz crystal microbalance (QCM) as both a substrate and mass accumulation measurement device during IBAD MgO film growth. Using the QCM apparatus and temporally correlated reflected high-energy electron diffraction (RHEED) imaging, we observed a sudden decrease in mass accumulation rate at the same time as the abrupt appearance of crystallographic texture[15]. In this paper, we further investigate the development of texture in MgO with the IBAD process using our in-situ QCM apparatus. We also develop the concept of control and manipulation of this model system to produce useful template layers for nanoscale applications by exploring the effect of the ion to molecule arrival ratio and the use of different substrates.

EXPERIMENT

Optically polished 50 nm gold-coated 14 mm diameter 5 MHz quartz crystals (< 3 nm rms) were purchased from Q-sense, Inc. (Glen Burnie, Maryland). These as-received crystals were placed in a custom made fixture. The fixture consisted of a water-cooled copper block that provided sufficient thermal contact with the oscillating crystal using a thin indium seal. The quartz crystal was used as both a QCM and substrate for subsequent depositions of a nucleation layer and the IBAD MgO film. A second integrated QCM crystal was used to monitor and control the MgO vapor flux. Additional films were deposited on silicon wafer substrates without the native oxide removed. In some cases, silicon nitride membranes (SPI, Inc.) were attached to these substrates for concurrent deposition of the IBAD MgO films.

All depositions were conducted in a high vacuum chamber with a typical base pressure of 7.0×10^{-6} Pa (5.0×10^{-8} torr) at room temperature. A four-pocket 7 cc Temescal SuperSource provided the deposit vapor flux. A two-grid collimated Kaufman ion source at an incidence angle of $45°$ relative to the substrate normal provided an Ar ion flux to the substrate. The ion fluence was monitored with a separate Faraday cup.

Several different thin films were deposited directly onto the in-situ Au-coated quartz crystal or Si substrate and used as IBAD MgO bed layers. Typically, a 10 to 20 nm thick amorphous layer of Si_3N_4 was used. No background N_2 gas was introduced into the chamber during the Si_3N_4 deposition. In some cases, a 10 to 20 nm thick layer of Y_2O_3 was deposited in a similar manner without the use of a partial pressure of oxygen.

The IBAD MgO layer was deposited with concurrent 750 eV Ar ions and MgO fluxes. An electron beam evaporator provided the MgO vapor flux at 0.05 nm/s to 0.10 nm/s. The flow rate of Ar gas into the system was kept constant at 10 sccm, which corresponded to a chamber pressure of ~$5.0{\times}10^{-3}$ Pa. The ion to atom ratio was varied to determine its effect on nucleation of the MgO films. IBAD film growth was monitored in-situ using RHEED. Frames were taken at 1-second intervals during deposition. The RHEED beam is aligned along an axis 90° relative to the ion beam. All patterns were taken at a beam energy of 25 keV.

Some samples were imaged in a Philips CM20 operating at an accelerating voltage of 200 kV. These samples were deposited on 3 mm diameter silicon nitride windows, as previously mentioned, to reduce sample preparation time and avoid the introduction of artifacts during ion milling. Plan view images were taken in bright field and dark field modes.

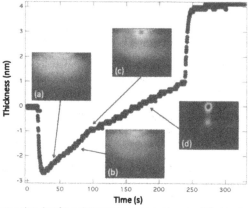

Figure 1. Thickness versus time data for an IBAD MgO film growth on a polished in-situ QCM crystal with correlated RHEED images. At position (a), a broad amorphous background is observed. As the film thickness increases, polycrystalline rings begin to form (b). Position (c) shows the first signs of texture formation. By time (d), the biaxial texture is fully formed.

RESULTS AND DISCUSSION

The use of the in-situ QCM apparatus coupled with correlated RHEED images allows us to observe and quantify aspects of the IBAD MgO growth. Figure 1 shows one such data set from a run with an ion to molecule ratio of ~0.40. When the QCM substrate is initially exposed to the ion beam and vapor flux, a transient is observed in the thickness data. This transient is recoverable once the shutter is closed at the end of the run. An initially diffuse background is observed by RHEED as the deposition begins (Figure 1(a)). As MgO begins to accumulate on the substrate, broad rings begin to form that correspond to the (200) and (220) MgO at Figure 1(b) indicative of a randomly oriented polycrystalline MgO film. Further deposition results in a point at which there is an abrupt change in slope. At this point (Figure 1(c)), we observe the onset of the spot pattern associated with the biaxial texture in MgO. The spot pattern continues to intensify with further deposition until higher order spots are visible at Figure 1(d).

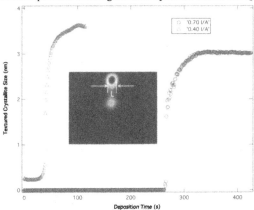

Figure 2. IBAD MgO crystallite size as a function of ion-to-molecule ratio. For a ratio of 0.40 ($\triangle$) the crystallite size increases to a value of ~3.5 nm. At the higher ratio of 0.70 (O), the crystallites reach a size of ~3 nm.

Data collected for IBAD experiments at different ion-to-molecule ratios exhibited a similar behavior[16]. In each case, as the ion-to-molecule ratio was increased, the change in slope occurred at ~ 2 nm film thickness, regardless of the ratio value. This suggests that there is a critical thickness for the development of the biaxially oriented MgO and is consistent with the island size of ~2 nm that survives ion impacts predicted in IBAD MgO simulations by Zepeda-Ruiz and Srolovitz[17]. These observations area also consistent with the model of a phase change during growth, similar to that advanced by Brewer and Atwater[14]. However, the presence of the broad polycrystalline MgO rings suggest that a grain selection process, similar to IBAD YSZ, may also be at work. Recent X-ray diffraction studies at the Stanford Synchrotron Radiation Laboratory show the definite presence of polycrystalline rings in both transmission and reflection geometries for thin (< 2 nm) IBAD MgO. These experiments corroborate our observations in the RHEED images during growth of IBAD MgO films. The intensity of the rings in the diffraction patterns also suggests that there are significant amounts of each orientation in the film prior to the onset of biaxial texture. These data warrant a more thorough analysis to determine the effect of this microstructure on the texturing of the film and will be reported at a later date. It should be noted that this data was taken at a single ion-to-molecule ratio of ~0.40 and we suspect that different ratios will change the distribution of crystallite orientations during growth.

The ion-to-molecule ratio can have an effect on the textured crystallite size. To investigate this effect in IBAD MgO, we have used a technique to determine the crystallite size by using calibrated RHEED images and measuring the full-width-at-half-maximum for a Gaussian fit through the low order (20) RHEED spot. Hartman et al.[18] have shown that RHEED images can be used to determine the in-plane and out-of-plane values and determine the lateral grain size of MgO films prepared by the IBAD process. We find that there is a difference in the textured crystallite size when different ion-to-molecule ratios (I/A) are used. Figure 2 shows that for two different I/As, the crystallite size differs by ~0.5 nm, being smaller for the higher I/A ratio. The crystallite size has been confirmed using high-resolution bright-field plan-view TEM imaging. Figure 3 shows the presence of lattice fringes that outline crystallites from ~3 to 5 nm in diameter and agree well with the data obtained from our RHEED images during growth.

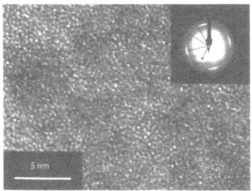

Figure 3. A high-resolution bright-field plan-view TEM image of an IBAD film deposited on a Si_3N_4 membrane window. The visible lattice fringes show small misorientations between crystallites that have a range of sizes from ~3 to 5 nm. The inset selected area diffraction pattern shows the texturing in this film as manifested in the arcs along each indexed ring.

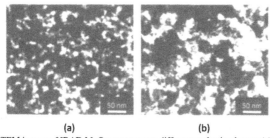

(a)   (b)

Figure 4. Darkfield TEM images of IBAD MgO grown on two different nucleation layers: (a) Si_3N_4 and (b) Y_2O_3. The MgO deposited on each layer have distinctly different morphologies and grain sizes. Both films are of comparable thickness. Figure 4(b) taken from ref. 19.

Another effect on the crystallite size is observed when using different nucleation layers for IBAD MgO. Figure 4 shows this effect with Si_3N_4 and Y_2O_3 (Figure 4(b)) layers. The dark-field image for the Si_3N_4 (Figure 4(a)) layer shows crystallite colonies with good grain-to-grain alignment on the order of 10 to 20 nm with small crystallites similar in size to those observed in the high-resolution image in Figure 3. Figure 4(b) shows that much larger colonies are present that contain larger crystallites for the Y_2O_3 case. Proper selection of the nucleation can influence the nucleation density and size of crystallites.

SUMMARY

We show data on the initial nucleation of biaxial texture in IBAD MgO using an in-situ quartz crystal microbalance (QCM) as a substrate combined with in-situ reflected high-energy electron diffraction (RHEED). Experimental data shows that the initially polycrystalline MgO film textures at a critical thickness of ~2 nm, regardless of the ion-to-molecule ratio, but RHEED images show the final crystallite size varies between 3 to 4 nm when the ratio is changed. Imaging with transmission electron microscopy has corroborated these observations. Changes in the ion-to-molecule ratio can be used to control the crystallite size. Growth of these films on various substrates changes the sticking coefficient of the MgO and influences the nucleation density and crystallite size as well. This suggests the possibility of using MgO and other materials to develop biaxially textured crystallites with a narrow, specified size distribution to be used on the nanoscale as templates or stand alone structures for such applications.

ACKNOWLEDGMENTS

The authors wish to thank the generous funding from both the Office of Electricity Delivery and Energy Reliability, U.S. Department of Energy and from the Center for Applied Superconductivity Technology of the 21st Century Frontier R&D Program funded by the Ministry of Science and Technology, Republic of Korea through a grant to Stanford University.

REFERENCES

1.   W. Ensinger, Surface and Coatings Technology **99** (1-2), 1-13 (1998).
2.   W. Ensinger, Surface and Coatings Technology **80** (1/2), 35 (1996).
3.   R. Hubler, A. Schroer, W. Ensinger, G. K. Wolf, W. H. Schreiner and I. J. R. Baumvol, Surface and Coatings Technology **60** (1/3), 561 (1993).
4.   L. S. Yu, J. M. E. Harper, J. J. Cuomo and D. A. Smith, Appl. Phys. Lett. **47** (9), 932 (1985).
5.   Y. Iijima, N. Tanabe, O. Kohno and Y. Ikeno, Appl. Phys. Lett. **60** (6), 769-771 (1992).
6.   P. N. Arendt and S. R. Foltyn, MRS Bulletin **29** (8), 543 (2004).
7.   C. P. Wang, K. B. Do, M. R. Beasley, T. H. Geballe and R. H. Hammond, Appl. Phys. Lett. **71** (20), 2955-2957 (1997).
8.   V. Selvamanickam, Y. Chen, X. Xiong, Y. Xie, X. Zhang, Y. Qiao, J. Reeves, A. Rar, R. Schmidt and K. Lenseth, Physica C: Superconductivity **463-465**, 482-487 (2007).
9.   R. M. Bradley, J. M. E. Harper and D. A. Smith, Journal of Applied Physics **60** (12), 4160 (1986).
10. R. Huhne, S. Fahler, B. Holzapfel, C. G. Oertel, L. Schultz and W. Skrotzki, Physica C **372/376** (SUPPL. 2), 825 (2002).
11. J. Dzick, J. Hoffmann, S. Sievers, L. O. Kautschor and H. C. Freyhardt, Physica C: Superconductivity **372-376** (Part 2), 723-728 (2002).
12. C. P. Wang, Doctor of Philosophy, Stanford University, 1999.
13. I. O. Usov, P. N. Arendt, J. R. Groves, L. Stan and R. F. DePaula, Nuclear Instruments & Methods in Physics Research, Section B (Beam Interactions with Materials and Atoms) **243** (1), 87 (2006).
14. R. T. Brewer and H. A. Atwater, Appl. Phys. Lett. **80** (18), 3388-3390 (2002).
15. J. R. Groves, R. H. Hammond, R. F. DePaula and B. C. Clemens, presented at the Fall 2008 MRS Meeting, Boston, MA, 2008 (in press).
16. J. R. Groves, R. F. DePaula, L. Stan, R. H. Hammond and B. C. Clemens, IEEE Trans. Appl. Supercond., in press (2009).
17. L. A. Zepeda-Ruiz and D. J. Srolovitz, Journal of Applied Physics **91** (12), 10169-10180 (2002).
18. J. W. Hartman, R. T. Brewer and H. A. Atwater, Journal of Applied Physics **92** (9), 5133-5139 (2002).
19. P. N. Arendt, High Temperature Superconductivity Program Peer Review, Washington, D.C. (2002).

Mater. Res. Soc. Symp. Proc. Vol. 1181 © 2009 Materials Research Society

Ion beam synthesis of Ge nanocrystals embedded in SiO$_2$ matrix

N Srinivasa Rao[1], A P Pathak[1*], D Kabiraj[2], S A Khan[2], B K Panigrahi[3], K G M Nair[3] and D K Avasthi[2]

[1]School of Physics, University of Hyderabad, Central University (P.O),

Hyderabad 500 046, INDIA

[2]Inter University Accelerator Centre, P.O.10502, New Delhi 110 067, INDIA

[3]Materials Science Division, Indira Gandhi Centre for Atomic Research,

Kalpakkam, 603 102, INDIA

ABSTRACT

High fluence of low energy Ge$^+$ ions were implanted into Si matrix. We have also deposited Ge and SiO$_2$ composite films using the Atom beam sputtering (ABS). The as implanted/as-deposited films were irradiated by Swift Heavy Ions (SHI) with various energies and fluences. These pristine and irradiated samples were subsequently characterized by XRD and Raman to understand the crystallization behavior. Raman studies of the films indicate the formation of Ge crystallites as a result of SHI irradiation. Glancing angle X-ray diffraction results also confirm the presence of Ge crystallites in the irradiated samples. Moreover, the crystalline nature of Ge improves with an increase in ion fluence. Rutherford back scattering was used to quantify the concentration of Ge in SiO$_2$ matrix and the film thickness. These detailed results have been discussed and compared with the ones available in literature. The basic mechanism for crystallization induced by SHI in these films will be presented.

*Corresponding author E-mail: appsp@uohyd.ernet.in & anandp5@yahoo.com
Tel: +91-40-23010181 / 23134316   Fax: +91-40-23010181 / 23010227.

INTRODUCTION

The structural, optical and electronic properties of low dimensional, indirect band gap materials have been investigated extensively over the past years. Semiconductor nanostructures of indirect band gap materials such as silicon (Si) and germanium (Ge) have been studied widely because of their potential applications in optoelectronics and nanophotonics [1, 2]. It has been suggested that direct optical transitions are possible in small size group-IV nanocrystals [3]. Although porous Si is expected to be the most promising Si-based light emitting material, nano crystalline Ge (nc-Ge) embedded in Silica matrices have their own advantages. Ge has smaller electron and hole effective masses and larger dielectric constant than Si. The effective Bohr radius of the exciton in Ge is larger than that in Si. Hence it is much easier to change the electronic structure around the band gap of Ge than Si due to its larger exciton Bohr radius [4]. Various methods have been used to prepare Ge nanocrystals embedded in various matrices. So far, a series of techniques have been used to prepare nc-Ge, including Sol-gel [5], Ion-Implantation [6], UV-assisted oxidation [7], and Co-Sputtering [8]. The Present work is to investigate the structure of the Ge nanoparticles grown by Atom Beam Sputtering (ABS) and recrystalization behavior of Ge implanted Si upon Swift Heavy Ion (SHI) irradiation. Atom Beam Sputtering (ABS) technique has been used to prepare metal nano particles embedded in silica matrix and found to be suitable for such studies [9]. The present work investigates the structure of the Ge nanoparticles grown by ABS. These irradiated samples were characterized by Raman, XRD and RBS.

EXPERIMENTAL DETAILS

Thin films of Ge and SiO_2 nanocomposites were prepared by ABS of Ge and silica with fast neutral Ar atoms having energy of ~ 1.5 keV. Thin films of the mixture of Ge and SiO_2 were deposited on Si and quartz substrates. We used atom source instead of an ion source in order to avoid the usual charge build up in ion beam sputtering. The depositions were carried out by co-sputtering of Ge and Silica using 1.5 keV Ar atoms. The Ar atom source was mounted at an angle of 45^0 facing the sputtering target inside a vacuum chamber, as shown in Fig.1.

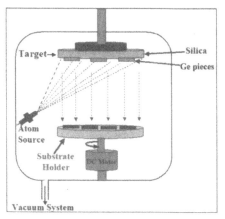

Fig.1: Schematic set-up of ABS

DC motor is used to rotate the substrate holder to get better uniformity. SHI irradiation was done at room temperature with 150 MeV Ag^{12+} ions with fluences varying from 2.5×10^{12} to 2×10^{13} ions /cm^2 from the 15 MV pelletron at IUAC. Low beam current was maintained to avoid heating of the samples during irradiation.

On the other hand 400 keV Ge^+ ions were implanted with a fluence of 5×10^{16} ions/cm^2 into a cleaned p-type Si(001) substrate. During implantation the substrate was maintained at 573 K. The as-implanted sample was irradiated with 100 MeV Au^{8+} ions with a fluence of 1×10^{13} ions/cm^2 to investigate the crystallization behavior of high fluence Ge implanted Si.

Raman scattering spectra of the films were obtained in backscattering configuration with a Renishaw Raman microscope using a 514.5 nm Ar+ laser excitation source. All the measurements were carried out at room temperature. X-ray diffraction measurements were carried out with CuKα X-rays, λ=0.154nm in a glancing angle incidence geometry. RBS was used for the quantification of Ge concentration in ABS films using He^{2+} ions.

RESULTS AND DISCUSSION

The Rutherford Backscattering Spectrum of the ABS pristine sample has been shown in Fig.2. The estimated Ge concentration in these films was 15%. This is in good agreement with the expected value. We found that the thickness of as-deposited film is around 100 nm.

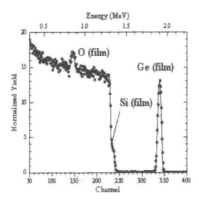

Fig.2: RBS spectra of Ge 15% as-deposited sample

The X-ray diffraction (XRD) spectra of the as-deposited and ion irradiated samples are shown in figure 3. In the case of as-deposited sample, no peak is observed, whereas the irradiated samples show various peaks of (113) and other planes, which indicates the formation of nc-Ge. With an increase in ion fluence, the XRD peak becomes sharper and the full width at half-maximum (FWHM) of peak decreases. It indicates that the average size of nc-Ge increases.

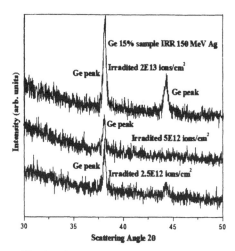

Fig. 3 GIXRD spectra of Ge 15% irradiated samples

The average size of the nanoparticles may be calculated using Scherrer's formula [10],

$$D = \frac{0.9\lambda}{\beta Cos\theta_B}$$

where λ = 1.5406A° (for Cu Kα), β = FWHM (in radians) and $2\theta_B$ is Bragg angle. The average diameter of nc-Ge is found to increase from 21 to 30 nm with an increase in ion fluence. When the ion fluence used for irradiation is increased, the mobility of Ge atoms and small nanocrystals increases. This leads to formation of Ge clusters and nanocrystals by diffusion of Ge atoms or small clusters inside the silicon oxide matrix. Therefore, increase in ion fluence resulted in increase of Ge crystallinity and particle size.

Fig. 4 shows the Raman spectra of the irradiated ABS samples with various fluencies. For sample irradiated with 2.5E12 ions/cm^2 we observed a broad hump around 280 cm^{-1}, indicating the presence of amorphous Ge. Such a band is typical of amorphous systems [11] where the Raman cross-section is proportional to the total density of vibrational states rather than to the spectral density of long-wavelength phonons [12].

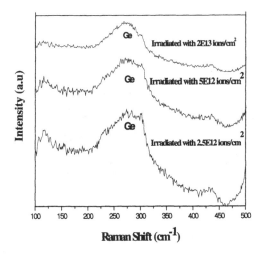

Fig.4   Raman spectra of Ge 15% irradiated samples

The presence of defects may also cause the increase in the width of the Ge peak. The peak position of Ge mode in this spectrum was shifted to lower wave number side. The broadness of the peak is reduced as a result of high fluence ion irradiation. The peak at higher fluence seems to be due to the presence of amorphous and crystalline nature of Ge. It is difficult to estimate the particle size from Raman spectra since the peak in the spectrum is due to combined presence of amorphous and crystalline nature of Ge.

X-ray diffraction spectra of pristine and irradiated samples of Ge implanted Si are shown in Fig. 5 at fixed fluence. This clearly indicates the formation of Ge nanoparticles. In Ge implanted Si due to high fluence the as-implanted sample shows the amorphous nature where as Au ion irradiation results in crystallization as evidenced from the Ge (311) at 53.746^0 and Si (311) at 55.84^0 peaks in the spectra. This is in accordance with the results obtained by implantation of Ge into thermally grown SiO_2 layer [13].

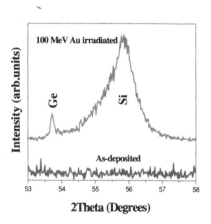

Fig. 5: GIXRD spectra of pristine and 100 MeV irradiated samples of Ge as-implanted Si

Under certain conditions Ge (111) or Ge (311) seems to be favorable. In our case we observed only the Ge (311) peaks. The appearance of Si peak indicates formation of the crystallized top layer. The large width of the Si peak implies existence of defects in the layer. The Raman spectrum of the Au ion irradiated sample is shown in Fig.6. Here the appearance of Ge mode along with SiGe mode indicates the recrystallization of the amorphized layer. Here the crystallization is not epitaxial. It is merely recrystallization of Ge implanted Si under SHI irradiation.

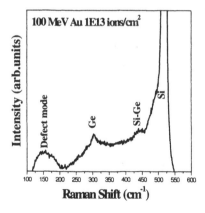

Fig. 6: Raman spectrum of irradiated Ge implanted Si sample with 100 MeV Au ions

Similar results have been reported for the annealed sample implanted with Ge ions [14]. The mode around 150 cm^{-1} may be attributed to residual damage in the top Si layer which is consistent with the broad Si peak in the X-ray spectra. We have chosen 100 MeV Au and 150 MeV Ag ions to study the electronic energy loss (S_e) dependence and the effect of S_e/S_n ratio in these samples to understand the recrystallization mechanism. One can explain the basic mechanism of recrystallization and formation of nanocrystals in these samples, prepared by two different methods involving ion-irradiation with the help of thermal spike model [15]. When the swift heavy ion passes through the material the energy is transferred to the target lattice via electron–phonon coupling. This large amount of energy is transferred to the target electrons in very short time, of the order of $\sim 10^{-12}$ seconds which leads to an increase of the lattice temperature above its melting point along the ion path. This results in a modification of structure of films around the cylindrical zone.

CONCLUSIONS

In conclusion, we have synthesized Ge nanocrystals embedded in SiO_2 matrix films by using swift heavy ion irradiation. The structural properties of the nanocrystals were characterized by XRD and Raman spectroscopy. The estimated diameter of nanocrystal is found to vary between 21 and 30 nm with an increase in ion fluence from 2.5×10^{12} ions/cm^2 to 2×10^{13} ions/cm^2. The Ge modes in the Raman spectra of irradiated samples indicate the presence of crystalline and amorphous nature of Ge. The variation of crystallite size as function ion fluence has been discussed. TEM investigations are under progress and will be reported later.

ACKNOWLEDGEMENTS

N.S.R. thanks CSIR, New Delhi for award of SRF. A.P.P. thanks IUAC for UFUP project. We would like to thank F. Singh and P Kulriya for Raman and XRD measurements. We acknowledge the help of S.R. Abhilash and D.C. Agarwal for the help during the experiments.

REFERENCES

1. Canham L T 1990 *Appl. Phys. Lett.* **57** 1046

2. Pavesi L *et al* 2000 *Nature* **408** 440

3. T Tanagahara and K Takeda, Phys Rev B 46, 15578 (1992)

4. Y Maeda, Phys. Rev. B 5, 1658 (1995).

5. M Nogami and Y Abe, Appl. Phys. Lett. 65, 2545 (1994).

6. Jia-Yu Zhang , Xi-Mao Bao and Yong-Hong Ye, Thin Solid Films 323, 68 (1998)

7. V Cracium, C B Leborgne, E J Nicholls and I W Boyd, Appl. Phys. Lett. 69, 1506 (1996).

8. A. V. Kolobov, S. Q. Wei, W. S. Yan, H. Oyanagi, Y. Maeda, and K. Tanaka Phys. Rev. B 67, 195314 (2003).

9. Y K Mishra, S Mohapatra, D Kabiraj, B Mohanta, N P Lalla, J C Pivin and D K Avasthi. Scripta Mater. 56, 629 (2007)

10. B. D. Cullity, Elements of X-ray Diffraction, 2nd. Ed., Addison-Wesley, reading, MA, 1978, p.102.

11. H Ritcher, Z P Wang, L Ley, Solid State Commun. 39 , 625 (1981)

12. W. Hayes, R. Loudon, Scattering of Light by crystals, Wiley, New York, 1978

13. P.K. Giri, R. Kesavamoorthy, S. Bhattacharya, B.K. Panigrahi, K.G.M. Nair. Materials Science and Engineering B 128, 201 (2006)

14. P K Giri, R Keasavamoorthy, B K Panigrahi and K G M Nair Nucl. Inst. and Meth. B 244, 56 (2006)

15. M. Toulemonde, C. Dufour, E. Paumier, Phys. Rev. B 46 (1992) 14362

Ion-Solid Interactions

Mater. Res. Soc. Symp. Proc. Vol. 1181 © 2009 Materials Research Society   1181-DD05-02

Quantum and Classical Molecular Dynamics Studies of the Threshold Displacement Energy in Si Bulk and Nanowire

E. Holmström[1], A. V. Krasheninnikov[1,2], and K. Nordlund[1]

[1] Helsinki Institute of Physics and Department of Physics, P.O. Box 43, FI-00014 University of Helsinki, Finland

[2] Department of Applied Physics, Helsinki University of Technology, P.O. Box 1100, FI-02015, Finland

ABSTRACT

Using quantum mechanical and classical molecular dynamics computer simulations, we study the full three-dimensional threshold displacement energy surface in Si. We show that the SIESTA density-functional theory method gives a minimum threshold energy of 13 eV that agrees very well with experiments, and predicts an average threshold displacement energy of 36 eV. Using the quantum mechanical result as a baseline, we discuss the reliability of the classical potentials with respect to their description of the threshold energies. We also examine the threshold energies for sputtering in a nanowire, and find that this threshold depends surprisingly strongly on which layer the atom is in.

INTRODUCTION

The threshold displacement energy of a material E_d is the single most fundamental quantity in determining the primary state of radiation damage in both bulk [1, 2] and nanoscale materials [3]. Knowing E_d in silicon is essential not only for the well-known use of the material in the manufacturing of semiconductor devices, but also because of contexts such as particle accelerators and space missions, where Si-based radiation detectors are exposed to extensive hadron damage. In spite of this vast technological interest and extensive scientific study of this quantity, the E_d averaged over all lattice directions remains poorly known in the material. Experimental methods show a widely varying scale of results for the average E_d in the range of 10 - 30 eV [4–7], and simulations using classical potentials show a similarly wide range of results [8–12].

The large variation of the simulation results on E_d is clearly related to the uncertainty of the interatomic potentials in the interaction energy range which determines the threshold. Since threshold displacement energies are typically in the range 10 - 50 eV [13–15], the part of the interatomic potential that determines the threshold is roughly speaking 1 - 20 eV above the minimum of the potential well. Unfortunately, no other commonly available experimental quantity depends on interaction energies in this range, and hence in this energy range the shape of the potential is usually just an extrapolation of a fit to much lower-energy data of elastic properties. In metals, in fact, potentials are commonly fit to known threshold displacement energies [16], but in Si this approach is not useful due to the large experimental uncertainties in the quantity.

Computer capacity has increased sufficiently to allow for dynamic simulation of atom motion in small systems using quantum mechanical approaches such as density-functional theory [17]. Since simulation of the threshold displacement energy only requires a few hundred atoms [18], it is thus natural to study E_d using fully quantum mechanical approaches to circumvent the problems associated with interatomic potential fitting. Until recently, however, such studies were limited to studying the threshold displacement energy only in a few principal directions [19–22], even though it is well known that the threshold displacement energy is in principle different in all non-equivalent lattice directions, and the average threshold is usually much larger than the values in the principal directions [15, 16].

We have recently systematically simulated the threshold displacement energy with DFT molecular dynamics (MD) methods in all lattice directions [23]. In this paper we review these results, and present a new comparison with classical potentials with the objective to determine which of three commonly used classical potentials (Stillinger-Weber [24], Tersoff [25] and EDIP [26, 27]) is the best for threshold energy simulations.

In addition to the bulk threshold displacement energy, also the threshold energy for sputtering at surfaces is of interest, particular in nanostructures, since these have a large area-to-volume ratio. Indeed, Si nanowires were recently irradiated with P and B ions[28], and contrary to what would happen in bulk Si for similar fluences, irradiation of Si nanowires was reported to induce a limited amount of amorphization and structural disorder, as proven by electrical transport and Raman measurements. Furthermore, the precise knowledge on the threshold energy is mandatory for using electron and ion beams for cutting and welding of Si nanowires [29]. Finally, a comparison of the results for Si nanowires with those obtained for other nanostructures, such as carbon nanotubes and onions, irradiated with energetic ions[30, 31] and electrons[32–34] should be of interest. Hence we have also examined the threshold displacement energy for sputtering (a form of damage production as a vacancy is left behind) in a Si nanowire with an axis in a $\langle 111 \rangle$ crystal direction.

SIMULATION METHODS

Bulk threshold displacement energy .

Determining E_d in a specific direction consists of simulating recoils in that direction with increasing energy until a permanent defect is obtained. In a previous paper we examined systematically the threshold energy in Fe, and reported that there are several possible definitions of both the minimum and average (effective) threshold displacement energies [16]. In the current paper we do not discuss these various definitions, but use the same notation for the quantities as in Ref. [16] to make it clear exactly what definition was used. All simulations were carried out at 0 K ambient temperature.

Obtaining a reliable value for $E_{d,ave}^{av}$ demands the determination of E_d in a number of randomly selected lattice directions, and therefore involves simulating a large number of individual recoil events in the lattice. Since DFT MD calculations are extremely heavy computationally, we used a *two-stage* approach to determine the thresholds.

In the *first stage*, we carried out classical MD simulations (which are many orders of magnitude faster) with decreasing system size and different temperature control schemes to find a simulation condition that allowed for a reliable determination of the threshold displacement energy with a minimum number of simulation time steps and atoms needed. From the comparison with larger system sizes and simulation times, we were also able to estimate the systematic error due to the simulation scheme used. This error is given with the subscript "SYST" in the results for the average DFT thresholds. These classical simulations were performed using the PARCAS [35] simulation code with the Tersoff [36] and Stillinger-Weber [37] potentials, and the DFT MD was done using SIESTA [38] as a force module to PARCAS.

The scaling tests showed that a system size of 144 atoms and a total simulation time of 3 ps (with no cooling during the first 200 fs, and a Berendsen cooling of all atoms after 200 fs with a time constant $\tau = 300$ fs [39]) allowed for determining the average threshold energy $E_{d,ave}^{av}$ with a systematic error of ± 1.7 eV [23]. The simulation cell was not cubic, but instead had the z-axis of the unit cell oriented in the $\langle 111 \rangle$ direction. This made it less likely that replacement collision sequences would interact with each other over the periodic boundaries.

In the *second stage*, we carried out scaling tests of the DFT simulation parameters to find the computationally lightest possible parameter sets that would still provide good formation energies of the basic types of point defects in the lattice, in both the local density approximation (LDA) and the generalized gradient approximation (GGA). The calculations of E_d within the GGA scheme, generally considered more reliable, provided confirmation for the calculations within the computationally lighter LDA scheme, which was used for the heavy task of determining $E_{d,ave}^{av}$. Defect formation energies were considered an obvious test for the system, as the defect formation energies are a fundamental and much studied quantity that are related to the threshold energy. The varied DFT parameters for the defect calculations were the basis set, the k-point sampling, and the equivalent plane-wave cutoff energy. The basis was varied between the single-zeta and double-zeta sets and each with polarization orbitals included, respectively. The number of k-points was scanned between a number of 18 and 4 points, and the single Γ point. The equivalent plane-wave cutoff energy was varied between 50.0 to 300.0 Ry.

The defect formation energy is defined as $E^f = (E_d/N_d - E_u/N_u)N_d$ where E_d and N_d are the potential energy and number of atoms in the defect cell, and E_u and N_u are the same in a defect-free cell [40]. For each tested set of parameters, the lattice constant of the cell was first relaxed. This was then followed by the relaxation of the structures of the dumbbell or split-$\langle 110 \rangle$ interstitial, the hexagonal interstitial, the tetrahedral interstitial, and the vacancy within a cell of the equilibrium lattice constant, using the conjugate gradient method. From these static calculations, it was found that only parameter sets utilizing the single-zeta basis could be considered in terms of computation time for the threshold energy MD simulations. Two such sets, one within the LDA and one within the GGA approximation, were chosen for the molecular dynamics runs to come. Four k-points, a cutoff energy of 100.0 Ry, and the Ceperley-Alder exchange-correlation functional were used for the LDA set, and four k-points, a cutoff energy of 250.0 Ry, and the Perdew-Burke-Ernzerhof exchange-correlation functional were used for the GGA set.

The formation energies of the defects calculated within these schemes as well as results from other DFT studies are presented in Table 1. Additionally, the ground state Frenkel pair, which was found to consist of a tetrahedral interstitial and a close vacancy, is included. In the LDA set, the hexagonal interstitial was not stable with respect to the relaxation, and instead relaxed towards a split configuration. There is considerable uncertainty about the interstitial formation energies. Our values are somewhat larger than those obtained in other studies, but the order in energy is similar [41–47]. In particular, most studies indicate that the dumbbell is the ground state of the isolated interstitial, in agreement with our result [42, 48].

Table 1: Formation energies of the basic point defects in eV calculated within the chosen LDA and GGA schemes, and results from other DFT studies. The ground state Frenkel pair for our calculations consists of a tetrahedral interstitial and a vacancy.

Defect	LDA	GGA	Other studies
Split-$\langle 110 \rangle$	4.7	4.9	2.88 - 3.84 [41–43, 48]
Hexagonal	-	5.6	2.87 - 3.80 [41–43, 48]
Tetrahedral	5.7	5.8	3.43 - 5.1 [41, 43]
Vacancy	3.4	3.9	3.17 - 3.65 [42, 43, 48, 49]
Frenkel pair	6.8	7.5	4.26 - 4.32 [48]
Bond defect	2.6	3.0	2.34 - 2.80 [42, 48]

The formation enthalpy of a Frenkel pair consisting of a tetrahedral interstitial and a vacancy was recently found to be lower than that of the corresponding pair with a dumbbell interstitial, the dumbbell however being the isolated ground state interstitial [50], which is well in line with our results.

Finally, as a separate test for the validity of the used Troullier-Martins pseudopotential at high interaction energies (relevant to close collisions between energetic atoms), the potential energy of the silicon dimer was computed for distances of $r = 0.80$ to 10 Å within the LDA and GGA schemes and compared to all-electron calculations [51]. The result was that reliable calculations could be performed up to energies of at least 100 eV, which easily suffices for calculations of E_d in any lattice direction in Si.

After the tests were finalized, E_d was calculated for a total of 80 random directions for atoms of both sublattices within the LDA scheme, and for 20 random directions within the GGA scheme. Additionally, E_d was calculated for the lattice directions $\langle 111 \rangle$ and $\langle 100 \rangle$. The simulation procedure was similar as in Ref. [16] except that the energy increment step was 4 eV.

Nanowire threshold displacement energy

For the calculations of the threshold energy for sputtering in nanowires, the Stillinger-Weber many-body empirical potential [37], which was found to overall give the best agreement with the DFT results, was used to model the Si-Si interactions.

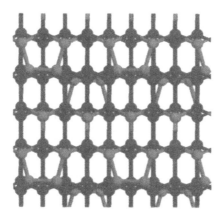

Figure 1: A close-up view of the reconstructed [112] surface of the wire. The reconstruction is 2 × (7.68 Å). The coloring indicates the potential energy of the atoms, where blue is low and green is high.

We created a nanowire in two steps. The wire was first cut out from a slab of silicon, then the surface reconstruction was carefully applied. Initially, a hexagonal rod was cut out from a bulk slab of silicon with the z-axis aligned in the $\langle 111 \rangle$ direction. The wire was 103 Å in length and 43 Å in diameter, each face of the cylinder displaying the [112] surface of silicon. Next, the reconstruction of the [112] surface was applied to the surface of the wire. This was the 2 × (7.68 Å) reconstruction that has been found by ab initio simulation as well as experiment [52, 53].

The initial positions for a unit cell of the reconstruction were found by first simulating the wire at a temperature of 300 K for 20 ps and then quenching the wire to 0 K at a rate of 0.01 K/fs. Periodic boundaries were applied in the z-direction of the unit cell in all simulations throughout this work. Due to the finite cutoff of the used potential, the surface was not orderly reconstructed, but parts of the surface displayed the periodically repeated unit cell of the desired reconstruction, as depicted in Fig. 1. Thus, the outermost layer of each of the six surfaces of the wire was cut out, and the unit cell of the reconstruction was duplicated onto the entire surface of the wire. After this, the wire was held again at a temperature of 300 K for 20 ps and then quenched to 0 K at a rate of 0.01 K/fs. Finally, conjugate gradient minimization was applied to the system.

The threshold displacement energies were subsequently determined with an approach similar to that used in the bulk with the classical PARCAS simulations, with the major difference that only events where an atom was sputtered were counted to be a defect. Thus these simulations give the threshold energy for sputtering. Although DFT-based simulations[32, 33] of displacement energies in carbon nanomaterials indicated that a Frenkel pair

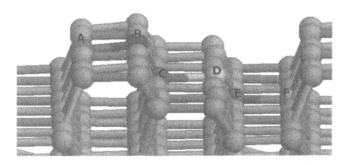

Figure 2: The six nonequivalent sublattices on the surface of a Si nanowire with a $\langle 111 \rangle$-oriented axis and all surface atoms reconstructed as $\langle 112 \rangle$.

can in principle be created when the sputtered atom is adsorbed on the surface at energies lower than the threshold for sputtering, the difference in energies is normally quite small, so one can assume that the displacement threshold for surface atons equals to the sputtering threshold. Each simulation covered 10 ps, and the system was cooled at the periodic boundaries using Berendsen temperature control with a time constant of $\tau = 300.0$ fs.

As an initial crosscheck against the bulk simulations, we determined the average threshold displacement energy for defect production for an atom in the center of the nanowire. This was found to be $29.1 \pm 0.8_{\text{STAT}}$ eV, which agrees within the uncertainties with the value of $28.95 \pm 0.15_{\text{STAT}}$ eV obtained in the bulk SW potential simulations (see above). The subscript "STAT" indicates that this is the statistical error of the average.

We then simulated the average threshold displacement energy for atoms on the surface of the nanowire. In doing this, one must consider that the reconstructed nanowire surface has six non-equivalent atom positions, as illustrated in Fig. 2. Hence all six positions were considered separately in the simulations.

RESULTS AND DISCUSSION

Bulk threshold energy

The results for the threshold displacement energies calculated within the LDA and GGA schemes are presented in Table 2. A large number of bond defects [43] were observed to be formed in the simulations, which somewhat complicated the determination of the global minimum as well as the average threshold energy. The results for the directions $\langle 111 \rangle$ and $\langle 100 \rangle$ given in the table are the threshold energies required for producing a defect of any type in those directions, either a Frenkel pair or a bond defect, and in these directions the lower threshold energy was for the Frenkel pair. However, bond defects were observed for a recoil energy as low as 12 eV in some random directions, and hence we also determined an average threshold energy $E_{d,ave}^{av,BD/FP}$ for producing *either* a bond defect *or* a Frenkel pair in addition to the average threshold energy for producing a Frenkel pair, $E_{d,ave}^{av,FP}$.

We obtained values of $36 \pm 2_{STAT} \pm 2_{SYST}$ and $35 \pm 4_{STAT} \pm 2_{SYST}$ eV for $E_{d,ave}^{av,FP}$ within the LDA and GGA schemes, respectively, and $24 \pm 1_{STAT} \pm 2_{SYST}$ and $23 \pm 2_{STAT} \pm 2_{SYST}$ eV for $E_{d,ave}^{av,BD/FP}$. The LDA and GGA values are sufficiently close for the LDA values, computed from a greater number of directions, to be considered statistically reliable.

It has been experimentally shown that Frenkel pair production in the $\langle 111 \rangle$ direction of the silicon lattice is greater than in the other two main lattice directions $\langle 100 \rangle$ and $\langle 110 \rangle$ [6, 54]. No value lower than 12.5 eV for E_d^{FP}, the threshold energy for producing Frenkel pairs, was obtained in any of the directions studied throughout the simulations, which is in full agreement with the experimental findings. Thus our results support the previous deduction that the global minimum of E_d^{FP} is in the open $\langle 111 \rangle$ direction and about 13 eV.

Table 2: Threshold displacement energies in eV calculated within the LDA and GGA schemes and classical Stillinger-Weber (SW), Tersoff and EDIP potentials. A and B denote the closed and open $\langle 111 \rangle$ directions, respectively. In the direction-dependent cases (top 3 rows), the uncertainty is a systematic uncertainty related to the choice of the energy increment step. For the average threshold (bottom 3 rows), the uncertainty is the standard error of the average (due to the averaging over many directions, the increment step size does not affect the average). The number of cases simulated for the averages was 80 for LDA, 20 for GGA, and 2500 for the classical potentials.

	LDA	GGA	SW	Tersoff	EDIP
$\langle 111 \rangle (A)$	14.5 ± 1.5	14.5 ± 1.5	20.5 ± 0.5	15.5 ± 0.5	15.5 ± 0.5
$\langle 111 \rangle (B)$	12.5 ± 1.5	12.5 ± 1.5	17.5 ± 0.5	10.5 ± 0.5	11.5 ± 0.5
$\langle 100 \rangle$	20.5 ± 1.5	19.5 ± 1.5	23.5 ± 0.5	9.5 ± 0.5	14.5 ± 0.5
$E_{d,ave}^{av,FP}$	36 ± 2	35 ± 4	28.95 ± 0.15	18.67 ± 0.10	16.23 ± 0.06
$E_{d,ave}^{av,BD/FP}$	24 ± 1	23 ± 2	20.08 ± 0.14	17.54 ± 0.08	13.31 ± 0.07

For both the GGA and LDA runs, the tetrahedral interstitial type accounts for about 75 % of all the Frenkel pair end state configurations, which is consistent with the finding that the ground state Frenkel pair includes a tetrahedral interstitial. The typical formation kinetics of this type of Frenkel pair invoke a replacement collision chain of the atoms in the $\langle 111 \rangle$ direction. The second type of observed end state interstitial is the dumbbell state, which was observed for 15% of all Frenkel pair end states. The remaining 10 % of Frenkel pair end states are characterized by a small cluster of vacancies and interstitials, instead of clearly separable point defects. In some 60% of all the directions studied, a bond defect complex was created with a lower threshold energy than a Frenkel pair. This is to be expected, as the formation energy of this defect was found to be considerably lower than that of the Frenkel pair.

The result of $12.5 \pm 1.5_{\mathrm{SYST}}$ eV for the global minimum of E_d^{FP} is in excellent agreement with the experimental value of 12.9 ± 0.6 eV obtained by Loferski and Rappaport for the onset of damage in silicon under electron irradiation [4]. Additionally, our result of $20 \pm 2_{\mathrm{SYST}}$ eV for E_d^{FP} in the $\langle 100 \rangle$ direction is in excellent agreement with the frequently quoted value of 21 eV from a study by Watkins and Corbett, where the orientation of the electron beam was along a $\langle 100 \rangle$ axis [5]. It should be noted, that the bond defect is invisible to standard experimental techniques [48], and therefore a comparison to experiment of our results concerning bond defects can not be made at the present. However, our observation of large numbers of bond defects is in excellent agreement with the very recent experimental deduction that the bond defect plays a significant role in the amorphization of silicon [55].

Values used for $E_{d,ave}^{av,FP}$ in the radiation effects and material damage community appear in the range of 13 to 25 eV [56–59]. The common usage of 13 and 15 eV for the value of the parameter is inappropriate, as from our simulations it is clear that the actual value is over a factor of two higher.

Table 2 also shows the results obtained with the classical potentials. These are also illustrated in Figure 3. The direction-specific thresholds agree well with previous literature values reported for the same potentials [20, 21]. Both the Tersoff and EDIP potentials underestimate the average and $\langle 100 \rangle$ direction thresholds, while the SW overestimates the thresholds in the $\langle 111 \rangle$ directions. Overall, the Stillinger-Weber potential is clearly closest to the DFT results, predicting both average and direction-dependent thresholds that compare reasonably well with the DFT results.

The contour plots in Figure 3 show that the overall behaviour of the threshold energies is similar in all potentials, with the smallest energies being in the $\langle 100 \rangle$ and $\langle 111 \rangle$ directions. Also, for all potentials, the highest thresholds are in directions surrounding the closed $\langle 111 \rangle$ direction. The maximal thresholds are in all cases in a circle of angles around closed $[111]$, at directions roughly $45°$ off $[111]$. E.g. the two maxima to the left and right of $[111]$ for the Tersoff potential are in roughly the $[\bar{1}44]$ and $[4\bar{1}4]$ directions

Nanowires

In the nanowires, we obtained average threshold displacement energies of 15.8 ± 0.5 eV (B position), 17.7 ± 0.9 eV (A), 17.8 ± 0.8 eV (D), 22.6 ± 0.9 eV (C), 28 ± 2 eV (E) and 29 ± 2 eV (F), where the values are listed from smallest to largest. Comparison of the order (B, A, D, C, E, F) with Fig. 2 shows that, not surprisingly, the high-lying atoms A and B are displaced the easiest, C and D the next-easiest, and E and F are most difficult to displace. That the displacement energy for the D atoms is almost as low as for A and B is clearly due to the fact that also atom D is on the outer edge of the step edge in the reconstruction. Naturally, atoms on the edges of the facets could have even lower displacement energies than the atoms on the facet sides, but these were not simulated separately.

What is somewhat unexpected is that comparison with the bulk average of 29.1 ± 0.8 eV shows the E and F atoms are practically as difficult to sputter as it is to displace an atom inside the bulk. This is surprising, since both atoms have movement directions where they could sputter without having to displace any other atom. This nicely illustrates that

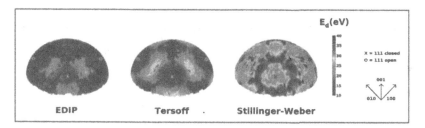

Figure 3: Threshold displacement energies for Frenkel pair production in the bulk in all lattice directions obtained with the EDIP, Tersoff and Stillinger-Weber classical potentials. The colors give the threshold energy in each direction. Values above 40 eV and below 10 eV are given with the same colour as for 40 eV or 10 eV, respectively.

the threshold displacement energy in Si is dominated by the need to break nearest-neighbor bonds.

The average threshold for all 6 different kinds of surface atoms is 22 ± 1 eV, i.e. clearly less than the bulk average. Since sputtering is always also associated with damage production (of at least one vacancy), this shows that damage creation in nanowires is clearly more pronounced on the wire surface.

CONCLUSIONS

Using SIESTA DFT molecular dynamics simulations, we have determined that the global minimum of the threshold displacement energy is $12.5 \pm 1.5_{SYST}$ eV, in the $\langle 111 \rangle$ direction, which is in excellent agreement with experimental results. The same simulations also gave the average threshold displacement energy for creating Frenkel pairs in silicon as $36 \pm 2_{STAT} \pm 2_{SYST}$ eV, which is clearly higher than the values used commonly in ion irradiation damage models. We find additionally that a bond defect complex is formed in most lattice directions with a lower threshold energy than a Frenkel pair. The threshold energy for forming either kind of defect is $24 \pm 1_{STAT} \pm 2_{SYST}$.

Out of the classical potentials examined (EDIP, Tersoff and Stillinger-Weber), we found that the Stillinger-Weber potential thresholds are clearly closest to the DFT results.

We also examined the threshold displacement energy on the surfaces of nanowires using the classical Stillinger-Weber potential. For $\langle 111 \rangle$-oriented Si nanowires with the $\langle 112 \rangle$ surface reconstruction, we found that the most outlying atoms have an average threshold displacement energy which is about a factor of 2 lower than atoms inside the nanowire or in the bulk. Somewhat surprisingly, however, already surface atoms on the bottom layer of the reconstruction had an average threshold energy which was almost equal to that in the bulk.

ACKNOWLEDGMENTS

We thank Dr. K. Oskenkorva for being a continued source of inspiration during the course of this work. This work was performed within the Finnish Centre of Excellence in Computational Molecular Science (CMS), financed by The Academy of Finland and the University of Helsinki. Grants of computer time from the Center for Scientific Computing in Espoo, Finland, are gratefully acknowledged.

REFERENCES

1. R. Smith (ed.), *Atomic & ion collisions in solids and at surfaces: theory, simulation and applications* (Cambridge University Prss, Cambridge, UK, 1997).

2. R. S. Averback and T. Diaz de la Rubia, in *Solid State Physics*, edited by H. Ehrenfest and F. Spaepen (Academic Press, New York, 1998), Vol. 51, pp. 281–402.

3. A. V. Krasheninnikov and F. Banhart, Nature Mater. **6**, 723 (2007).

4. J. Loferski and P. Rappaport, Phys. Rev. **111**, 432 (1958).

5. J. Corbett and G. D. Watkins, Phys. Rev. **138**, 555 (1965).

6. P. Hemment and P. Stevens, J. Appl. Phys. **40**, 4893 (1969).

7. D. Marton, in *Low Energy Ion-Surface Interactions*, edited by J. W. Rabalais (Wiley, Chester, 1994), p. 526.

8. L. Miller, D. Brice, A. Prinja, and S. Picraux, Phys. Rev. B. **49**, 16953 (1994).

9. L. Miller, D. Brice, A. Prinja, and T. Pricraux, in *Defects in Materials, MRS Symposia Proceedings No. 209*, edited by P. D. Bristowe, I. E. Epperson, I. E. Griffith, and Z. Liliental-Weber (Materials Research Society, Pittsburgh, 1991), p. 171.

10. M.-J. Caturla, T. Diaz de la Rubia, and G. H. Gilmer, in *Materials Synthesis and Processing Using Ion Beams, MRS Symposia Proceedings No. 316*, edited by R. I. Culbertson, O. W. H. amd K. S. Jones, and K. Maex (Materials Research Society, Pittsburgh, 1994), p. 141.

11. L. Miller, D. Brice, A. Prinja, and T. Pricraux, Radiat. Eff. Defects Solids **129**, 127 (1994).

12. M. Sayed, J. H. Jefferson, A. B. Walker, and A. G. Cullis, Nucl. Instrum. Methods Phys. Res. B **102**, 232 (1995).

13. P. Lucasson, in *Fundamental Aspects of Radiation Damage in Metals*, edited by M. T. Robinson and F. N. Young Jr. (ORNL, Springfield, 1975), pp. 42–65.

14. H. H. Andersen, Appl. Phys. **18**, 131 (1979).

15. P. Jung, Phys. Rev. B **23**, 664 (1981).

16. K. Nordlund, J. Wallenius, and L. Malerba, Nucl. Instr. Meth. Phys. Res. B **246**, 322 (2005).

17. For a review, see R. O. Jones and O. Gunnarsson, Rev. Mod. Phys. **61**, 689 (1989).

18. J. B. Gibson, A. N. Goland, M. Milgram, and G. H. Vineyard, Phys. Rev **120**, 1229 (1960).

19. S. Uhlmann *et al.*, Radiat. Eff. Defects Solids **141**, 185 (1997).

20. W. Windl, T. J. Lenosky, J. D. Kress, and A. F. Voter, Nucl. Instr. and Meth. B **141**, 61 (1998).

21. M. Mazzarolo, L. Colombo, G. Lulli, and E. Albertazzi, Phys. Rev. B **63**, 195207 (2001).

22. A. V. Krasheninnikov, Y. Miyamoto, and D. Tománek, Phys. Rev. Lett. **99**, 016104 (2007).

23. E. Holmström, A. Kuronen, and K. Nordlund, Phys. Rev. B **78**, 045202 (2008).

24. F. H. Stillinger and T. A. Weber, Phys. Rev. B **31**, 5262 (1985).

25. J. Tersoff, Phys. Rev. B **38**, 9902 (1988).

26. M. Z. Bazant, E. Kaxiras, and J. F. Justo, , (1997).

27. J. F. Justo *et al.*, Phys. Rev. B **58**, 2539 (1998).

28. A. Colli *et al.*, Nano Letters **8**, 2188 (2008).

29. S. Xu *et al.*, Small **1**, 1221 (2005).

30. A. V. Krasheninnikov, K. Nordlund, and J. Keinonen, Appl. Phys. Lett. **81**, 1101 (2002).

31. J. A. Åström, A. V. Krasheninnikov, and K. Nordlund, Phys. Rev. Lett. **93**, 215503 (2004).

32. A. V. Krasheninnikov *et al.*, Phys. Rev. B **72**, 125428 (2005).

33. T. Loponen, A. V. Krasheninnikov, M. Kaukonen, and R. M. Nieminen, Phys. Rev. B **74**, 073409 (2006).

34. L. Sun *et al.*, Phys. Rev. Lett. **101**, 156101 (2008).

35. K. Nordlund, 2006, PARCAS computer code. The main principles of the molecular dynamics algorithms are presented in [60, 61]. The adaptive time step and electronic stopping algorithms are the same as in [62].

36. J. Tersoff, Phys. Rev. B **38**, 9902 (1988).

37. F. A. Stillinger and T. A. Weber, Phys. Rev. B **31**, 5262 (1985).

38. J. M. Sole *et al.*, J. Phys.: Condens. Matter **14**, 2745 (2002).

39. H. J. C. Berendsen *et al.*, J. Chem. Phys. **81**, 3684 (1984).

40. G.-X. Qian, R. M. Martin, and D. J. Chadi, Phys. Rev. B **38**, 7849 (1988).

41. W.-K. Leung *et al.*, Phys. Rev. Lett. **83**, 2351 (1999).

42. O. K. Al-Mushadani and R. J. Needs, Phys. Rev. B **68**, 235205 (2003).

43. M. Tang, L. Colombo, J. Zhu, and T. Diaz de la Rubia, Phys. Rev. B **55**, 14279 (1997).

44. G. A. Baraff and M. Schluter, Phys. Rev. B **30**, 3460 (1984).

45. Y. Bar-Yam and J. D. Joannopolous, Phys. Rev. Lett. **52**, 1129 (1984).

46. R. Car, P. J. Kelly, A. Oshiyama, and S. T. Pantelides, Phys. Rev. Lett. **52**, 1814 (1984).

47. R. Car, P. J. Kelly, A. Oshiyama, and S. T. Pantelides, Phys. Rev. Lett. **54**, 360 (1985).

48. S. Goedecker, T. Deutsch, and L. Billard, Phys. Rev. Lett. **88**, 235501 (2002).

49. M. J. Puska, S. P.öykkö, M. Pesola, and R. M. Nieminen, Phys. Rev. B **58**, 1318 (1998).

50. S. A. Centoni *et al.*, Phys. Rev. B **72**, 195206 (2005).

51. K. Nordlund, N. Runeberg, and D. Sundholm, Nucl. Instr. Meth. Phys. Res. B **132**, 45 (1997).

52. C. H. Grein, J. Crys. Growth **180**, 54 (1997).

53. C. Fulk *et al.*, J. Electr. Mat. **35**, 1449 (2006).

54. V. S. Vavilov, V. M. Patskevich, B. Y. Yurkov, and P. Y. Glazunov, Fiz. Tverd Tela **2**, 1431 (1960).

55. P. D. Edmondson, D. Riley, R. C. Birtcher, and S. E. Donnelly, (2008), to be published.

56. G. P. Summers, E. A. Burke, and R. J. Walters, IEEE Trans. in Nucl. Sci. **40**, 1372 (1993).

57. J. F. Ziegler, SRIM-2003 software package, available online at *http://www.srim.org*.

58. R. E. MacFarlane, RSIC, PSR-118 / NJOY (1979).

59. M. Huhtinen, Nucl. Instrum. Methods Phys. Res. A **491**, 194 (2002).

60. K. Nordlund *et al.*, Phys. Rev. B **57**, 7556 (1998).

61. M. Ghaly, K. Nordlund, and R. S. Averback, Phil. Mag. A **79**, 795 (1999).

62. K. Nordlund, Comput. Mater. Sci. **3**, 448 (1995).

Mater. Res. Soc. Symp. Proc. Vol. 1181 © 2009 Materials Research Society   1181-DD13-06

Study of Silicon Carbide Ceramics

Malek Abunaemeh[1], Ibidapo Ojo[1], Mohamed Seif[2], Claudiu Muntele[1] and Daryush ILA[1]
[1]Center for Irradiation of Materials, Alabama A&M University, Normal, AL 35762
[2]Mechanical Engineering Department, Alabama A&M University, Normal, AL 35762

Abstract

The TRISO fuel that is intended to be used for the generation IV nuclear reactor design consists of a fuel kernel of Uranium Oxide (UOx) coated in several layers of materials with different functions. One consideration for some of these layers is Silicon Carbide (SiC) [1]. The design, manufacture and fabrication of SiC are done at the Center for Irradiation of Materials (CIM). This light weight material can maintain dimensional and chemical stability in adverse environments and very high temperatures. The characterization of the elemental makeup of the SiC material used is done using X-ray photoelectron spectroscopy (XPS). Nano-indentation is used to determine the hardness, stiffness and Young's Modulus of the material. Raman Spectroscopy is used to characterize the chemical bonding for different sample preparation temperatures.

Introduction

Tristructural-isotropic (TRISO) fuel particles were originally developed in Germany for high-temperature gas-cooled reactors[1].The first nuclear reactor to use TRISO fuels was the AVR, a prototype pebble bed reactor at Jülich Research Centre in West Germany, and the first power plant was the THTR-300, a thorium high-temperature nuclear reactor rated at 300 MW electric. TRISO fuels are also being used in experimental reactors such as the HTR-10 in China and the HTTR in Japan.

TRISO is a type of micro fuel particle. It consists of a fuel kernel composed of Uranium dioxide (UO$_X$) [1,2] in the center, coated with four layers of three isotropic materials. The four layers are a porous buffer layer made of carbon, followed by a dense inner layer of pyrolytic carbon (PyC), followed by a ceramic layer of SiC to retain fusion products at elevated temperatures and to give the TRISO particle more structural integrity, followed by a dense outer layer of PyC. TRISO fuel particles are designed not to crack due to the stresses from different processes (such as differential thermal expansion or fusion gas pressure) at temperatures beyond 1600°C. Therefore it can contain the fuel in the worst accident scenarios in a properly designed reactor. Two such reactor designs are the pebble bed reactor (PBR), in which thousands of TRISO fuel particles are dispersed into graphite pebbles, and the prismatic-block gas-cooled reactor (such as the GT-MHR), in which the TRISO fuel particles are fabricated into compacts particles[2] and placed in a graphite block matrix. Both of these reactor designs are very high temperature reactors (VHTR) [formally known as the high-temperature gas-cooled reactors (HTGR) [2], one of the six classes of reactor designs in the Generation IV initiative.

The TRISO fuel is directly immersed in the cooling fluid that extracts the heat outside of the reactor core while keeping the inside within the operational temperature limits. If the fissile fuel is in direct contact with the cooling fluid, there are great chances that radioactive fission fragments will be carried out of the reactor core and contaminate all other equipment. Therefore, in order to minimize such leaks, the fuel is by design coated with some diffusion barrier materials, SiC being one such choice [3].

Silicon Carbide is a compound mixture of Silicon and Carbon. This material can maintain dimensional and chemical stability in adverse environment and very high temperatures (up to 3000°C) [3]. It is also impermeable to gases and chemically inert. There are several forms in which SiC exist from coatings to whiskers to powders. There are different methods that are used to produce SiC. The easiest method is to combine the silica sand and carbon in an Acheson graphite oven and heat the mixture to a temperature between 1800 and 2600 °C. Another method that is used to manufacture SiC is the Poco method [4]. This method involves taking the pre fabricated, purified and processed graphite and chemically invert it to SiC. SiC is found in a solid physical state [5]. This material should be handled with care following good industrial hygiene practice. SiC is also non-flammable with a weak explosion hazard [6].

The main purpose of this work is to understand the changes in fundamental properties (chemical and mechanical stability) of SiC ceramics after exposure to temperatures up to 2500°C. This study will help to determine its eligibility for future irradiation testing for a specific application in an extreme radiation environment in the nuclear reactor. It will also help to determine if SiC will be a good choice as a diffusion barrier in the fuel cells of the TRISO fuel that will be used in the next generation of nuclear reactors.

Methods and Instrumentations

We took Phenolic resin and mixed it with Silicon powder. This Silicon powder used has a grain size of 10 μm. The ratio was determined to be 8.5g of Si nano-powder to 100g of PF resin. The mixture is then placed in a sonication bath for a few hours, then poured in the desired mold that is wrapped with Aluminum foil to prevent the material from sticking in the mold during the demolding process, and heated slowly at a rate of 20°C/day to lower the possibility of any bubble forming during the gelling stage. The hardened gel is removed from the mold and Aluminum foil slowly, not to break the sample, then heated to 2500°C slowly over the course of five days. The heating profile for glassy polymeric carbon (GPC) was followed during this process [7]. The sample is then left to cool off in the furnace until it reaches room temperature. Figure 1 shows a plot of the heating profile used for the sample.

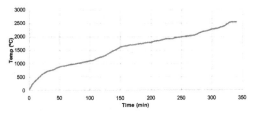

Figure 1: Heating profile used for the SiC samples in graphite furnace

Several methods and instruments were used to characterize the elemental and mechanical properties for our SiC samples. The first one was the X-ray photoelectron spectroscopy (XPS). XPS is a surface chemical analysis technique used to identify the elemental and chemical components that make up the sample. The sample is irradiated with a beam of X-rays. The number of electrons at each kinetic energy (KE) value is then counted. Al K_α Raman spectroscopy is another technique that was also used to characterize the changes in the chemical bonding for different temperatures under which our samples were prepared. A He-Ne laser

(632nm) was used. A G200 model nano-indenter was used for mechanical characterization of the material. This nano-indenter is equipped with a four-sided pyramid diamond Vickers indenter which makes it perfect to apply to almost any contact geometry. Information from the nano-indentation was used to calculate the hardness, stiffness and Young's Modulus.

Results and discussion:

XPS data were looked at for the SiC sample after 2500°C where SiC should form. At 1000°C we did not have the need to do XPS or nano indentation as that was mostly to verify from Raman that no SiC was formed at that temperature. We noticed an instrumental energy shift of ~ 4 eV that we need to take into consideration when looking at the XPS data, and that we did not correct for in displaying the results, therefore all values shown are literature values plus 4 eV. Figure 2a shows the XPS spectra for the SiC samples after heating it to 2000°C. Figure 2b shows the XPS spectra for the SiC samples after heating it to 2500°C. Both graphs (figures 2a and 2b) shows where sp3 Carbon lines that are associated with SiC and imperfections in graphitic Carbon, and the sp2 Carbon lines that are associated with the graphite were observed.

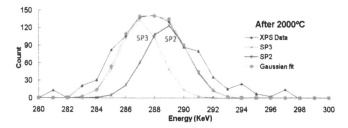

Figure 2a: XPS Spectrum for the C lines in the sample after heating to 2000°C

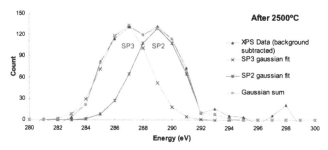

Figure 2b: XPS Spectrum for the C lines in the sample after heating to 2500°C

XPS data of a side cut of the sample shows more traces of sp2 Carbon formed than sp3 Carbon as seen in Figure 3. This leads to the conclusion that less SiC form inside the sample, most likely due to inhomogeneous distribution and sedimentation of the original Si powder in the Phenolic resin.

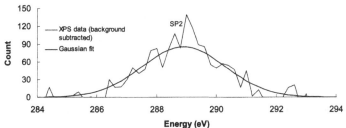

Figure 3: XPS spectrum showing mostly sp2 Carbon inside the sample.

From XPS we also observed that SiO₂ was formed on the sample as seen in Figure 4, attributed to original Aluminum oxide contamination of the sample from the Aluminum liner of the mold. During the heating, the aluminum oxide decomposed, leading to formation of Aluminum Silicide and Silicon Dioxide, both observed in XPS (Figures 4 and 5)

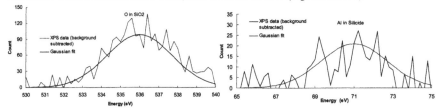

Figure 4 (left): XPS spectra of SiO₂ found in the sample
Figure 5 (right): XPS spectra of Aluminum Silicide found in the sample

Raman Spectroscopy was used to monitor the changes in the chemical bonding for Si and C constituents at various temperatures; Figure 6 shows the Distorted (D) and Graphitic (G) lines for the sample after various temperatures. The ratio of D/G, as seen in table 1 is increasing.

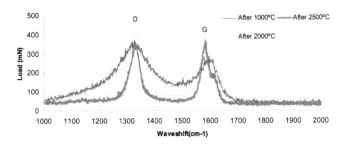

Figure 6: Raman Spectroscopy showing the C peaks at various temperatures

	After 1000°C	After 2000°C	After 2500°C
D	358	450	359
G	355	382	270
D/G	1.01	1.18	1.33

Table 1: The ratio of the Distorted over Graphitic in Raman

The Raman spectra shown in Figures 7 and 8 present peaks characteristic to 6H-SiC[8] formed in our sample. Nano-Indentation was also done to determine the hardness of SiC. A set of continuous load-displacement data is shown in Figure 9. Some important quantities are the peak load and displacement, the residual depth after unloading, h_f, and the slope of the initial portion of the unloading curve. The slope would have the dimensions of force per unit distance. Table 2 shows the calculated hardness, stiffness and Young's Modulus from the nano-indentation done.

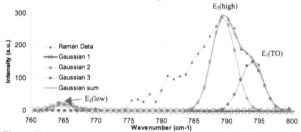

Figure 7: Raman Spectroscopy of SiC prepared at 2000 degrees C

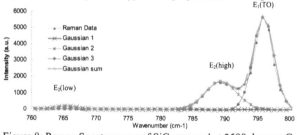

Figure 8: Raman Spectroscopy of SiC prepared at 2500 degrees C

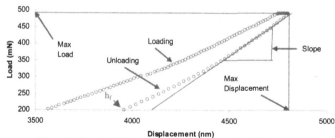

Figure 9: Detailed Nano Indentation Spectra for SiC

	After 2500°C
Stiffness (Slope) (kN/m)	700
Hardness (VPN)	777
Young's Modulus (GPa)	62.5

Table 2: mechanical calculation for SiC Sample (VPN is Vickers pyramid number)

Conclusions

The Raman results show evidence of 6H-SiC formation in the sample. Future X-ray diffraction or electron diffraction measurements will be needed in order to verify the Raman results. There were also traces of unreacted Graphite and Silicon from the Resin precursor, as well as Aluminum Silicide and Silicon Dioxide from contamination with Aluminum Oxide from the Aluminum liner of the mold, as indicated by the XPS spectra. This leads to the conclusion that the samples needed to be held at 2500°C for a longer period of time, and greater care needs to be taken in sample preparation to minimize contamination. The nano indentation values obtained are about an order of magnitude less than those of bulk 6H-SiC from the literature, suggesting that the grains of Silicon Carbide were not bonded among themselves. The next step is to prepare new samples but extending the time period where the SiC is heated to 2500°C. The same tests will be run again and the results will be compared.

Acknowledgments

This research was supported and funded by the AAMRI Center for Irradiation of Materials, NSF Alabama GRSP EPSCOR, and DoE NERI-C project number DE-FG07-07ID14894.

References

1. B.F. Myers, F.C. Montgomery and K.E. Partain, "The Transport of Fission Products in SiC," Doc No. 909055, GA Technologies Inc., (1986).
2. Kugeler and Schulten,1989, Application of boron and gadolinium burnable poison particles in UO2 and PUO2 fuels in HTRs
3. P. Krautwasser, G. M. Begun, and Peter Angelini, *J. of American Ceramic Society*, **66** 424 (1983).
4. Properties and characteristics of Silicon carbide, http://www.poco.com, Poco Graphite Inc
5. Madar, Roland, Aug 2004, "Materials science: Silicon carbide in contention". *Nature* **430** (430): 974–975. doi:10.1038/430974a
6. http://www.alfa.com/
7. Jonathan Fisher, Dec 1996, "Active Crucible Bridgman system: Crucible characterization, RF Inductor Design And Model", Dissertation, Alabama A&M University
8. Bhatnagar, M.; Baliga, B.J., March 1993, "Comparison of 6H-SiC, 3C-SiC, and Si for power devices". *IEEE Transactions on Electron Devices* **40** (3): 645–655. doi:10.1109/16.199372

Mater. Res. Soc. Symp. Proc. Vol. 1181 © 2009 Materials Research Society   1181-DD13-15

Low-energy ion beam sputtering of pre-patterned fused silica surfaces

J. Völlner, B. Ziberi, F. Frost, B. Rauschenbach

Leibniz-Institut für Oberflächenmodifizierung e.V. (IOM)
Permoserstr. 15, D-04318 Leipzig, Germany

E-mail: jens.voellner@iom-leipzig.de

ABSTRACT

Ripple formation and smoothing of pre-patterned fused silica surfaces by low-energy ion beam erosion have been investigated. As pre-pattern ripple surfaces produced by low-energy Ar^+ ion beam erosion were used. In addition to the enhanced ripple formation on the pre-patterned surfaces also the smoothing characteristics of surface is changed. Due to the anisotropic surface roughness of the ripple pattern the irradiation direction with respect to the pre-pattern becomes important. It is suggested that all of these effects are related to surface gradient dependent sputtering and therefore it is an important mechanisms also in the low-energy ion beam erosion of fused silica surfaces.

INTRODUCTION

In 1962, Navez *et al.* observed the first ripple structures on glass surface induced by ion beam erosion [1]. Since then a diversity of surface pattern topographies such as well ordered dot and ripple structures were found for nearly all material classes and have been discussed extensively. Different roughening and smoothing processes were discussed which contribute to surface patterning and smoothing by energetic ion beams [2-4].
Nevertheless, the experimental and theoretical work in the area of self-organized ion beam patterning of glasses such as fused silica is less investigated. As a potential optical material with notable physical and chemical properties fused silica is nearly irreplaceable as a material for lenses or light deliveries especially in the deep ultraviolet spectral range (DUV).
The evolution of surface features on fused silica [5] and glass [6] was investigated depending on, e. g. the ion incidence angle, ion energy or fluence. Although, the Bradley-Harper model [7] predicts the ripple orientation successfully, the ripple coarsening over time is not explained yet. In general, former investigations of the evolution of fused silica during ion beam erosion were focused on initially smooth surfaces. Recently it was shown for Si and Ag surfaces that the initial surface topography can strongly affects the ion induced ripple formation [8,9].
Therefore, in the present study surface smoothing and ripple formation on pre-patterned fused silica surfaces were investigated. As pre-pattern ripple surfaces produced by low-energy Ar^+ ion beam erosion at oblique ion incidence angles were used. The rippled surface also offers a model system to study the influence on a non-isotropic surface roughness on the ripple formation and surface smoothing.

EXPERIMENTAL METHODS

For the experiments described below a custom-built ion beam etching system with a base pressure below 2×10^{-6} mbar was used. The samples were mounted on a water-cooled sample holder that offers the possibility of rotating around its axis with about 12 rotations per minute. Additionally, it can be tilted from 0° up to 90° in steps of one degree, where the given ion incidence angles (α_{ion}) refers to the angle spanned by the surface normal and the axis of the ion-beam source. For ion beam sputtering a home-build Kaufman-type ion-beam source supplied with a two-grid ion optical system was used. Low-energy ion beam erosion was performed with Ar^+ ions (working pressure of 7×10^{-5} mbar) and the ion current density j_{ion} was kept constant at about 300 μA cm^{-2} corresponding to an ion flux of 1.87×10^{15} cm^{-2} s^{-1}. Samples used in this work were commercially available semi-polished fused silica wafers from HOYA Corporation with a rms (root mean square) roughness R_q of (0.4 nm ± 0.1 nm). For the investigation of the surface topography mainly scanning force microscopy (AFM) were applied.

RESULTS AND DISCUSSIONS

Surface evolution on plan surfaces

In Figure 1 different characteristics for the evolution of fused silica surfaces under low-energy Ar^+ ion beam erosion (E_{ion} = 1200 eV) are summarized briefly (see also [5]). In detail, Figure 1 shows the ion incidence angle dependent surface evolution quantified by the rms roughness R_q. Basically it is found that, depending on the incidence angle, different surface topographies emerge. For ion incidence angles less than a critical angle, surface smoothing is observed (marked yellow). For larger ion incidence angles (approx. 45° for the given example), ripple structures start to develop with a characteristic wavelength λ (marked blue). At angles ≥ 45° the evolving ripple patterns are aligned orthogonal to the ion beam and achieve their maximum in roughness, i. e. highest ripple amplitudes at 60°. An AFM image of a typical ripple pattern formed at 45° ion incidence is also shown. Near grazing incidence ≥ 70° ion beam erosion produce ripple structures with a parallel alignment with respect to the ion beam projection but with reduced amplitudes (rms roughness). Whereas for ion incidence angles of 50° the wavelength of the ripple structures increases for extended sputter times, the ripple

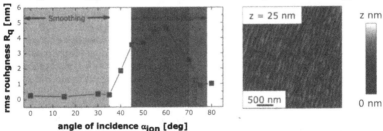

Figure 1. An overview of fused silica surface topographies by low-energy ion beam erosion (Ar^+, E_{ion} = 1200 eV, j_{ion} = 300 μA cm^{-2}, t = 60 min). Topography regimes and alignment of the ripples patterns are marked. The AFM image shows a characteristic ripple pattern as used for the experiments, the direction of the ion beam is indicated with the blue arrow.

wavelength saturates at ion incidence angles of 60° and 70°. The coarsening of the ripple pattern with increasing erosion time is illustrated in Figure 2 (E_{ion} = 1200 eV, α_{ion} = 50°). In this example the irradiation time was varied from 1 min to 120 min corresponding to total fluences of 1.1×10^{17} cm^{-2} to 1.3×10^{19} cm^{-2}. In the following set of experiments surfaces with different ripple wavelength and amplitudes, according to the erosion time, were used in order to investigate the influence of the pre-pattern on the smoothing as well as on the ripple formation on non-planar surfaces.

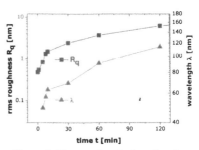

Figure 2. Temporal evolution of surface roughness R_q and ripple wavelength λ at α_{ion} = 50° (E_{ion} = 1200 eV, j_{ion} = 300 μA cm^{-2})

Formation and rotation of ripples on pre-patterned surfaces

In a first step a ripple pre-pattern was formed with an ion energy of 2000 eV, erosion time 120 min, and an ion incidence angle of 50° resulting in a ripple pattern with a characteristic ripple wave vector parallel to ion beam projection, a ripple wavelength of $\lambda \sim 180$ nm and a rms surface roughness of ~ 11.4 nm. Afterwards the sample was rotated azimuthally by 90° and irradiated again at an ion incidence angle of 50°, i. e. parallel to the ripples of the original pre-pattern, and for different erosion times. Fig. 3 (a,e) show the initial ripple pre-pattern. The images [Fig. 3 (b,f) – (d,h)] represent the temporal evolution of the surface during the irradiation parallel to the initial ripple pre-pattern. For all AFM image the corresponding FFT images are also shown [Fig. 3 (e) – (h)]. After 5 min of irradiation, it is clearly seen [Fig. 3 (b,f)] that new

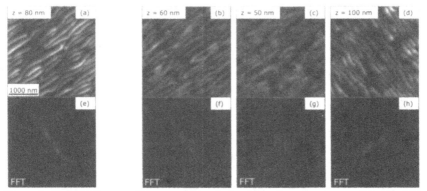

Figure 3. AFM/FFT images of the temporal evolution for ion beam erosion parallel to the initial ripple pre-pattern (α_{ion} = 50°). (a) initial pre-pattern and for (b) 5 min, (c) 10 min, (d) 60 min erosion time. The size of the images is 3×3 μm^2. The blue arrow indicates the ion beam direction. 3 (e) – (h) are the corresponding 2D-FFT images.

131

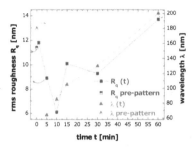

Figure 4. Temporal evolution of surface roughness R_q and ripple wavelength λ for the parallel irradiated pre-patterns.

ripples develop oriented orthogonal to the ion beam projection and also orthogonal to the former ripples of the pre-pattern. The emerging new ripples increase in wavelength with erosion time [Fig. 3 (c), (d): 10 min]. Simultaneously the original pre-pattern structure disappears. After 60 min of irradiation only ripples with an orthogonal orientation to the current ion beam projection are visible [Fig. 3 (d)]. Fig. 4 summarizes the temporal evolution of the rms roughness and ripple wavelength of the new ripples as obtained from an analysis of the AFM images. After a transient time of smoothing ($\sim$ 10 min) the orthogonal orientated ripple pattern coarsen continuously in wavelength and amplitude with increasing erosion time. A comparison of the initial pre-pattern (erosion time 120 min, $R_q = 11.4$ nm, $\lambda \sim 180$ nm) with the new developed ripple pattern (erosion time 60, $R_q = 13.7$ nm, $\lambda_{60\,min} \sim 200$ nm) demonstrates that the ripple formation is enhanced on pre-patterned, non-planar, surfaces confirming investigation for the influence of surface roughness on the formation on ripples on Si [8].

Smoothing of pre-patterned surfaces

For the second set of experiments two rippled pre-patterns with different surface roughness values (and ripple wavelength) were produced by changing the erosion time. For sample #1 the surface roughness amounts to $R_q = 6.3$ nm, whereas for sample #2 the roughness was $R_q = 9.4$ nm. Both samples have been eroded at an ion incidence angle of 20° where the projection of the ion beam was, in the first case, parallel and, in the second case, orthogonal to

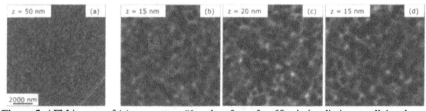

Figure 5. AFM images of (a) pre-pattern #1 and surface after 60 min irradiation parallel and orthogonal to the ripples of the pre-patterns [(b),(c)] and for simultaneous rotation (d). The image size is 10×10 μm². The blue arrows indicate the ion beam direction. The blue circle indicates sample rotation.

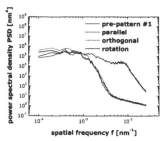

Figure. 6. PSD curves for the pre-pattern #1 and after erosion parallel and orthogonal to the pre-patterns and with simultaneous sample rotation.

the initial present ripple pattern. Additionally, experiments with a simultaneous sample rotation (12 rpm) during erosion were conducted. According to Fig. 1 for all cases a surface smoothing is expected. Fig. 5 shows AFM images of the pre-patterned fused silica surface #1 [Fig. 5 (a)] and after erosion parallel and orthogonal to the ripples of the pre-patterns [Fig. 5 (b) – (c)] and for the case of simultaneous rotation [Fig. 5 (d)]. All samples were irradiated with Ar$^+$ (E_{ion} = 2000 eV, α_{ion} = 20°, 60 min). The AFM images indicate a smoothing of the rippled pre-pattern where the surface roughness decreases from 6.3 nm to ~ 2 nm independent of the irradiation setup (parallel, orthogonal, rotation). This is also confirmed by a PSD analysis of the AFM images shown in Fig. 6. The diagram emphasizes a smoothing for all spatial frequencies > 10^{-3} nm^{-1} (spatial wavelength < 1 µm). In contrast, for sample #2 surface smoothing was only observed for the parallel erosion setup and in the case of a simultaneous sample rotation as seen from Fig. 7. For the orthogonal erosion [Fig. 7 (c)] hole-like structures, similar to cylinder sections, develop resulting in a dramatically enhanced surface roughness. The power spectral density functions for the different irradiations setups are summarized in Fig. 8. The smoothing for the parallel irradiated and the rotated sample for all spatial frequencies > 10^{-3} nm^{-1} is evident. For the hole-like structure (orthogonal irradiation) smoothing only occurs at wavelengths below ~290 nm. The similarity of the depressions formed on the orthogonal irradiated sample with surfaces features found for the erosion of fused silica and Si at higher ion energies [10,11] suggest that surface gradient dependent sputtering is an important mechanisms also in the low-energy ion beam erosion of fused silica surface.

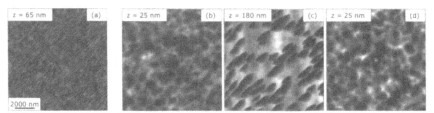

Figure 7. AFM images of (a) pre-pattern #2 and surface after 60 min irradiation parallel and orthogonal to the ripples of the pre-patterns [(b),(c)] and for simultaneous rotation (d). The image size is 10 × 10 µm^2. The blue arrows indicate the ion beam direction. The blue circle indicates sample rotation.

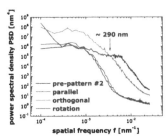

Figure 8. PSD curves for the pre-pattern #2 and after erosion parallel and orthogonal to the pre-patterns and with simultaneous sample rotation.

CONCLUSIONS AND OUTLOOK

In this contribution the ripple formation and smoothing of pre-patterned fused silica surfaces by low-energy ion beam erosion have been investigated. It was found that the ripple formation is amplified on pre-patterned surface. The initially present surface roughness also influences the smoothing characteristics of surface. In addition, for anisotropic surface roughness (ripple pattern) the irradiation direction with respect to the pre-pattern is important. It is suggested that all of these effects are related to surface gradient dependent sputtering [12] and therefore it is an important mechanisms also in the low-energy ion beam erosion of fused silica surface [13].

ACKNOWLEDGEMENTS

This work is supported by the Deutsche Forschungsgemeinschaft (FOR 845)

REFERENCES

1.   M. Navez, C. Stella, D. Chaperot, *C. R. Acad. Sci.* Paris **254**, 240 (1962)
2.   G. Carter, *J. Phys. D: Appl. Phys.* **34**, p 1 (2001)
3.   U. Valbusa et al., *J. Phys.: Condens. Matter* **14**, 1853 (2002)
4.   W. L. Chan, E. Chason, *J. Appl. Phys.* **101**, 121301 (2007)
5.   D. Flamm, F. Frost, D. Hirsch, *Appl. Surf. Sci.* **179**, 96 (2001)
6.   A. Toma et al., *Nucl. Instrum. Methods B* **230**, 551 (2005)
7.   R. M. Bradley, J. M. E. Harper, *J. Vac. Sci. Technol. A* **6**, 2390 (1988)
8.   P. Karmakar et al., *Appl. Phys. Lett.* **93**, 103102 (2008)
9.   A. Toma et. al., *J. Appl. Phys.* **104**, 104313 (2008)
10.   T. Motohiro, T. Taga, *Thin Solid Films* **147**, 153 (1987)
11.   G. W. Lewis et al., *Nucl. Instrum. Methods* **170**, 363 (1980)
12.   G. Carter, J. S. Collignon, M. J. Nobes, *J. Mater. Sci.* **8**, 1473 (1973)
13.   J. Völlner, B. Ziberi, F. Frost, B. Rauschenbach, to be published

Mater. Res. Soc. Symp. Proc. Vol. 1181 © 2009 Materials Research Society 1181-DD13-25

Sputtering yield measurements with size-selected gas cluster ion beams

Kazuya Ichiki[1], Satoshi Ninomiya[2], Toshio Seki[1,4], Takaaki Aoki[3,4], Jiro Matsuo[1,4]
[1] Department of Nuclear Engineering, Kyoto University, Sakyo, Kyoto, Japan
Fax: 81-774-38-3978, e-mail: ichiki.kazuya@nucleng.kyoto-u.ac.jp
[2] Quantum Science and Engineering Center, Kyoto University, Uji, Kyoto, Japan
[3] Department of Electronic Science and Engineering, Kyoto University, Nishikyo, Kyoto, Japan
[4] CREST, Japan Science and Technology Agency (JST), Chiyoda, Tokyo, Japan

ABSTRACT

Ar cluster ions in the size range 1000–16000 atoms/cluster were irradiated onto Si substrates at incident energies of 10 and 20 keV and the sputtering yields were measured. Incident cluster ions were size-selected by using the time-of-flight (TOF) method. The sputtering yield was calculated from the sputtered Si volume and irradiation dose. It was found that the sputtering yields decreased with increasing incident cluster size under the same incident energy conditions. The integrated sputtering yields calculated from the sputtering yields measured for each size of Ar cluster ions, as well as the cluster size distributions, were in good agreement with experimental results obtained with nonselected Ar cluster ion beams.

INTRODUCTION

When a cluster ion strikes a target surface, each constituent atom hits the local area at the same time and multiple-collision processes occur. It was found that the irradiation effects of cluster ions exceeded the sum of the individual irradiation effects of constituent atoms. For example, it was studied that the sputtering yield induced by a dimer was higher than twice that induced by the atomic projectile at the same velocity [1]. Under small (<10 atoms/ion) cluster ion bombardment [2, 3], these nonlinear effects were reported to be experimentally a function of cluster size. In contrast, there were only a few reports investigating the nonlinear effects under large (>100 atoms/cluster) cluster ion bombardment. We have investigated the irradiation effects of the large gas cluster ion beam (GCIB) [4]. Gas cluster is an aggregate of several thousand atoms and each constituent atom of 10 keV gas cluster has an energy of only a few eV. Many unique phenomena, such as high sputtering yield and surface smoothing under large gas cluster ion bombardment, have been observed [5]. Therefore, GCIB technology is expected to be a powerful tool for surface modification and analysis. In molecular dynamics simulation studies, it was reported that although 20 keV Ar_{2000} impact penetrates the Si surface and creates crater-like damage on the target, impact with 20 keV Ar_{10000} does not even penetrate the Si surface [6, 7], indicating that incident cluster size and energy are the important factors of large cluster irradiation effects such as damage formation and sputtering. Nevertheless, there are only a few reports about the relationship between incident cluster size and irradiation effects in experimental studies with large gas cluster ion beam, because the GCIB size distribution extends over more than several thousand atoms and it is very difficult to measure experimentally the irradiation effects of a cluster of specific size without size selection. Irradiation effects have been studied with the size-selected GCIB by using a strong magnet [8, 9], but the magnetic field intensity required for bending a large cluster ion is proportional to the square root of the incident ion size and energy. The maximum cluster size that can be separated with the magnetic field is about tens of thousands atoms/cluster. Secondary ion mass spectrometry (SIMS) with size-selected GCIB was also studied by using the TOF method [10, 11]. However, the GCIB current intensity with size selection decreases to less than 1/1000 of before size selection. Therefore, this method has been used only for SIMS experiments, where the required ion current is very low. In this study, the primary ion beam was focused and the current density of the continuous cluster ion beam was increased to more than 100 μA/cm² with an electrostatic lens to solve this problem and the sputtering yield of Si with size-selected GCIB was investigated.

EXPERIMENT

Fig. 1 shows a schematic view of the gas cluster ion irradiation equipment. Neutral Ar cluster beams were formed by supersonic expansion through a nozzle, and were ionized by electrons with energy in the range 70–300 eV emitted from a hot tungsten filament. The mean cluster size of the GCIB was roughly controlled by the inlet source gas pressure, ionization voltage and emission current. The ionized clusters were accelerated to energies from 10 to 20 keV. Primary ion beam size selection was performed by the TOF method using the two pairs of ion deflectors installed along the beam line. First, the Ar cluster ion beam was chopped to a width of 5 μs by applying a high-voltage pulse at the first deflector. The pulsed ion beam at the first deflector had the same energy and contained various sizes of cluster ions because of the size distribution of the clusters generated, and therefore small cluster ions would reach the second deflector before larger ones. The pulsed ion beam was chopped again to a width of 5 μs at the second deflector after an appropriate time delay from the first pulse. The pulsed ion beam at the second deflector contained a specific size of cluster ions depending on the delay time (t_D) between the two pulses. The flight length between the first and second deflectors was about 370 mm. The selected Ar cluster size N was proportional to V_a and $t_D{}^2$, where V_a is the acceleration voltage. The electrostatic einzel lens installed in front of the second deflector was used for focusing the ion beam to a 1 mm spot on the target. The primary ion beam was incident on the target at an angle of 0° with respect to the surface normal. The irradiation beam current was measured with a

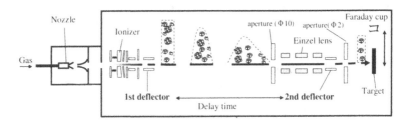

Fig. 1. The experimental setup for the size-selected GCIB irradiation system. Primary cluster ions were size-selected by using the TOF method. The primary cluster ion beam was chopped at the first deflector and chopped again at the second deflector after an appropriate delay time from the first pulse. The size-selected cluster ion beam was focused into a 1 mm spot on the target at a current density of higher than 50 nA/cm².

Faraday cup. The base and working pressures in the irradiation chamber were 1×10^{-5} and 1×10^{-4} Pa, respectively. The repetition frequency was 5000 Hz and the current intensity was from 0.5 to 5 nA after size selection. The etching volume was measured *ex situ* by an interferometric surface profiler (Maxim-NT, Zygo, USA). The observation area was 2.6 mm × 2.4 mm and the spatial resolution was 10 μm (X, Y axis) and 0.1 nm (Z axis) respectively.

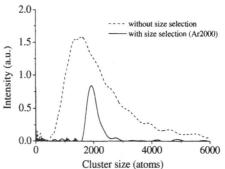

Fig. 2. Cluster size distributions of the nonselected (dashed line) and size-selected (full line) Ar cluster ions. The size range was from 500 to 5000 atoms/cluster before size selection and from 2000 ± 500 atoms/cluster after size selection, respectively.

RESULT AND DISCUSSION

The incident Ar cluster ion size was selected in the range of 1000–16000 atoms/cluster in this study. Before size selection, the mean cluster size was roughly controlled to use cluster ions efficiently. For example, the incident cluster ions of Ar_{2000} and Ar_{16000} were separated from the gas cluster ion beam with the mean size of about 2000 and 7000 atoms/cluster, respectively. The dashed line in Fig. 2 presents the size distribution of the nonselected cluster ion beam with mean size of about 2000 atoms/cluster, and the full line presents that of the size-selected cluster ion

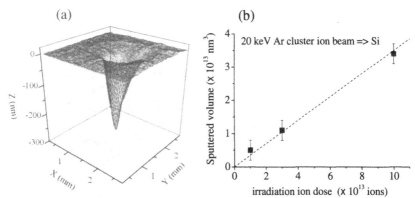

Fig. 3. (a) The surface profile of Si measured by the interferometric surface profiler after irradiation of 20 keV nonselected Ar cluster ion beam with the mean size of 2000 atoms/cluster at the dose of 3×10^{13} ions.
(b) Variation of sputtered volume of Si with irradiation dose of 20 keV Ar cluster ions. The sputtered volume was proportional to the irradiation dose.

beam (Ar$_{2000}$). The widths of the distribution functions of the nonselected and size-selected cluster ion beams were about 500 to 5000 and 2000±500 atoms/cluster, respectively. Both the size distribution and ion beam current decreased with shortening the pulse width. It was difficult to provide enough current of the smaller cluster than Ar$_{1000}$ in this experimental setup. The current density of the nonselected cluster ion beam was higher than 100 μA/cm^2. The maximum current density of the size-selected cluster ion beam was about 500 nA/cm^2 for Ar$_{1000}$ irradiation, and current density was maintained at about 50 nA/cm^2 for Ar$_{16000}$ irradiation.

Fig.3 (a) shows an example of surface profile for the Si surface irradiated with 20 keV nonselected Ar cluster ions. The profile was measured with the interferometric surface profiler. The mean size of irradiated cluster ions was 2000 atoms/cluster and the irradiation dose was 3 × 10^{13} ions. In this case, to save irradiation time, the cluster ion beam was not scanned. The sputtered volume was calculated directly from the surface profile because of the high depth resolution. The sputtered depth of Si irradiated with the nonselected GCIB with scanning measured by the interferometric surface profiler and by a contact surface profiler (Dektak3, Veeco, New York, USA) agreed well. Fig. 3 (b) shows the variation of sputtered volume of Si with an irradiation dose of a 20 keV Ar cluster ion beam. The sputtered volume was proportional to the irradiation dose, indicating that the sputtering yields can be evaluated by the surface profiles and Ar cluster ion dose even if the irradiation dose is as small as 1 × 10^{13} ions. In irradiation experiments for size-selected Ar cluster ions, each dose was higher than 5 × 10^{13} ions.

Fig. 4 shows the effects of incident cluster size on sputtering yields for 10 and 20 keV size-selected Ar cluster ions. The horizontal error bars represent the size width of size-selected Ar cluster ion beam. Under bombardment with 20 keV Ar$_{1000}$, more than 50 Si atoms were sputtered, and this sputtering yield was about 30 times higher than with 20 keV Ar atomic ions. The sputtering yields decreased with increasing cluster size because of the decreasing incident energy

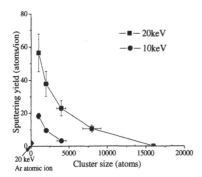

Fig. 4. The effects of incident Ar cluster size on Si sputtering yield for 10 and 20 keV Ar cluster ions. Sputtering yields decreased with increasing size when the total energy was the same.

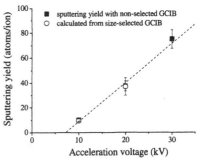

Fig. 5. The sputtering yields of Si with 10 to 30 keV Ar cluster ion. Solid squares show the sputtering yields measured with nonselected Ar cluster ion beams. The open circles were calculated from the incident size distribution and the sputtering yields measured with size-selected Ar cluster ions. Calculated sputtering yields agreed with the sputtering yields of nonselected cluster ion beams.

of each constituent atom. However, the sputtering yield of 20 keV Ar_{8000} was maintained about six times higher than that with 20 keV Ar atomic ions. The energy of each constituent atom for 20 keV Ar_{8000} was about 2.5 eV, which is lower than the value of the surface binding energy of Si (about 4.6 eV [12]). It is supposed to be the result of the effect of multiple collisions between the cluster ion and surface atoms.

Fig. 5 shows the sputtering yields of Si with Ar cluster ions. The solid squares present the sputtering yields with 10 to 30 keV nonselected Ar cluster ions and the mean size of incident cluster ions was 2000 atoms/cluster. The sputtering yields of Si with Ar clusters increased linearly with acceleration voltage. These sputtering yields (Y) for cluster ion beams containing various sizes of clusters can be represented as follows:

$$Y = \frac{\int Y(n)I(n)dn}{\int I(n)dn} \tag{1}$$

where n is the number of constituent atoms, $I(n)$ is the beam intensity of the n-size cluster ions and $Y(n)$ is the sputtering yields for the n-size cluster ions. Open circles in Fig. 5 present the calculated sputtering yields from the equation (1). As clearly shown in Fig. 5, the calculated yields agreed well with the yields for nonselected Ar cluster ions. It indicates that the irradiation effects under the incident of clusters of varying size can be reproduced by the integral of those for each size of cluster ions. It also shows that the cluster size selection with the TOF method is useful for the investigating irradiation effects with large cluster ions.

CONCLUSIONS

The sputtering yields of Si were measured for bombardment with Ar cluster ion of various sizes. Incident cluster ion sizes were selected by using the TOF method. The sputtering yields decreased with increasing incident Ar cluster size under conditions of same total energy. The sputtering yields with cluster ion beams containing various sizes of clusters were in good agreement with the sputtering yields calculated from the sputtering yields with each size of cluster ions and cluster size distributions, indicating that the irradiation effects with cluster ion beams having very broad size distributions can be reproduced by the integral of those for each size of cluster ions. The TOF method for incident size selection is very useful to investigate the irradiation effects of large cluster ions.

ACKNOWLEDGMENTS

This work is supported by the Core Research for Evolutional Science and Technology (CREST) of Japan Science and Technology Agency (JST). It is also supported in part by the Research Fellowships of the Japan Society for the Promotion of Science (JSPS) for Young Scientists.

REFERENCES

1. H. H. Andersen and H. L. Bay, *J. Appl. Phys.* 46, 2416–2422 (1975)
2. M. Benguerba, A. Brunelle, S. Della-Negra, J. Depauw, H. Joret, Y. Le Beyec, M. G. Blain, E. A. Schweikert, G. Ben Assayag and P. Sudraud, *Nucl. Instr. Meth.,* **B 62**, 8–22 (1991)

3. H. H. Anderson, A. Brunelle, S. Della-Negra, J. Depauw, D. Jacquet and Y. Le Beyec, *Phys. Rev. Lett.*, **80,** 5433–5436 (1998)

4. I. Yamada, J. Matsuo and N. Toyoda, A. Kirkpatrick, *Mater. Sci. Eng.,* **R 34,** 231–95 (2001).

5. J. Song, D. Choi and W. Choi, *Nucl. Instr. Meth.,* **B 196,** 275–78 (2002).

6. L. Rzeznik, B. Czerwinski, B. Garrison, N. Winograd and Z. Postawa, *J. Phys. Chem.,* **C 112,** 521–31 (2008).

7. T. Aoki, J. Matsuo and G. Takaoka, *Nucl. Instr. Meth.,* **B 202,** 278–82 (2003).

8. N. Toyoda, S. Houzumi and I. Yamada, *Nucl. Instr. Meth.,* **B 242,** 466–68 (2006).

9. K. Nakamura, S. Houzumi, N. Toyoda, K. Mochiji, T. Mitamura and I. Yamada, *Nucl. Instr. Meth.,* **B 261,** 660–63 (2007).

10. K. Moritani, M. Hashinokuchi, J. Nakagawa, T. Kashiwagi, N. Toyoda and K. Mochiji, *Appl. Sur. Sci,* **255,** 948–950 (2008).

11. S. Ninomiya, Y. Nakata, Y. Honda, K. Ichiki, T. Seki, T. Aoki and J. Matsuo, *Appl. Sur. Sci,* **255,**1588–1590 (2008).

12. Y. Yamamura and H. Tawara, Atomic Data and Nuclear Data Tables, **62,** 149–253 (1996)

Biological and Biomedical
Applications

Mater. Res. Soc. Symp. Proc. Vol. 1181 © 2009 Materials Research Society   1181-DD08-01

Fabrication and Surface Modification of Porous Nano-Structured NiTi Orthopedic Scaffolds for Bone Implants

Shuilin Wu[1, 2], Xiangmei Liu[1, 2], Paul K. Chu[1,*], Tao Hu[1], Kelvin W. K. Yeung[1], and Jonathan C. Y. Chung[1]

[1] Department of Physics & Material Science, City University of Hong Kong, Tat Chee Avenue, Kowloon, Hong Kong

[2] Ministry-of-Education Key Laboratory for the Green Preparation and Application of Functional Materials, School of Materials Science and Engineering, Hubei University, Wuhan 430062, China

*E-mail: paul.chu@cityu.edu.hk

ABSTRACT

Near-equiatomic porous nickel-titanium shape memory alloys (NiTi SMAs) are becoming one of the most promising biomaterials in bone implants because of their unique advantages over currently used biomaterials. For example, they have good mechanical properties and lower Young's modulus relative to dense NiTi, Ti, and Ti-based alloys. Porous NiTi SMAs are relatively easy to machine compared to porous ceramics such as hydroxyapatite and calcium phosphate that tend to exhibit brittle failure. The porous structure with interconnecting open pores can also allow tissue in-growth and favors bone osseointegration. In addition, porous NiTi alloys remain exhibiting good shape memory effect (SME) and superelasticity (SE) similar to dense NiTi alloys. To optimize porous NiTi SMAs in bone implant applications, the current research focuses on the fabrication methods and surface modification techniques in order to obtain adjustable bone-like structures with good mechanical properties, excellent superelasticity, as well as bioactive passivation on the entire exposed surface areas to block nickel ion leaching and enhance the surface biological activity. This invited paper describes progress in the fabrication of the porous materials and our recent work on surface nanorization of porous NiTi scaffolds in bone grafts applications.

INTRODUCTION

As a type of biometals, near-equiatomic nickel-titanium shape memory alloys (NiTi SMAs) have achieved great commercial success in the biomedical industry due to their unique shape memory effect (SME), superelasticity (SE), as well as good biocompatibility[1]. However, the higher Young's modulus and the dense structure limit the application of these alloys to bone implants because the mismatch in the Young's moduli between the implants and the substitutive tissues does not favor load transfer in the loading state. Consequently, formation of new tissues is adversely affected. To overcome these shortcomings, three-dimensional (3D) porous NiTi SMAs scaffolds have been proposed for tissue repair or reconstruction. In addition to possessing the desirable SME and SE properties intrinsic to the dense materials, porous NiTi scaffolds have porous structures and interconnective open channels that allow mass transport, cell migration, attachment, and proliferation, as well as tissues ingrowth. A number of methods have thus been developed to prepare porous NiTi scaffolds[2,3,4].

In recent years, the surface characteristics of biomaterials such as topography and roughness are increasingly becoming recognized as crucial factors with respect to tissue acceptance and cell behavior,[5,6,7] and some recent reports reveal that a nanoscale topography significantly influences the adhesion and proliferation of many types of cells such as osteoblast[8] and human mesenchymal stem cells[9] as well as genomic response[10]. Therefore, ideal tissue scaffolds should have hierarchical porous structures, implying that the scaffolds should possess not only macro 3D porous structures on the micro/millimeter scale to permit transport of nutrition and tissue ingrowth, but also surface features on the nanometer scale to provide a bioactive surface environment for cell or tissue growth. In addition, collagen and hydroxyapatite (HAP), the main constituents of bone, consist of arranged arrays of tropocollagen molecules (300 nm long and 1.5 nm wide) and needle-like HAP crystals (40–60 nm long, 20 nm wide and 1.5–5 nm thick).[11] Therefore, surface modification is necessary in order to produce a nano-structured surface on porous NiTi scaffolds. Because of the 3D porous structure and complex surface morphology, planar nano-patterning techniques are not applicable to surface nanorization of porous NiTi scaffolds at present, and in fact not many papers in this area can be found. In our previous publication, we reported the formation of a large number of nanowires and nanobelts on the surface of porous NiTi scaffolds using a facile hydrothermal treatment[12]. In this work, we describe the fabrication processes and present our recent investigation on the surface nanorization of porous NiTi scaffolds as well as subsequent effects on the biological properties.

THEORETICAL FOUNDATION AND DEVELOPMENT OF PM PROCESSES FOR POROUS NITI

As one of metal-based scaffolds in hard tissue engineering, porous NiTi alloys are usually fabricated by powder metallurgy (PM). Only the NiTi phase exhibits the unique SME and SE properties and secondary phases such as Ni_3Ti and $NiTi_2$ can only impair these properties. According to the Ni-Ti phase diagram that NiTi is the predominant phase when the atomic ratio of Ni and Ti is near 1:1 at a sintering temperature of about 1118 °C[13], equiatomic nickel and titanium elemental powders are typically used as the starting materials. With the exception of using pre-alloyed NiTi powders as the raw materials, all the PM fabrication processes of porous NiTi scaffold reported so far are based on the following exothermic reactions consisting of elemental Ni and Ti powders as the starting materials:

$$Ni + Ti \rightarrow NiTi + 67 \text{ kJ/mol} \tag{1}$$

$$Ni + Ti \rightarrow Ti_2Ni + 83 \text{ kJ/mol} \tag{2}$$

$$Ni + Ti \rightarrow Ni_3Ti + 140 \text{ kJ/mol} \tag{3}$$

Although porous NiTi made by PM methods is based on the same exothermic reactions between Ni and Ti, the different fabrication processes result in big differences in these products, particularly the mechanical properties and porous structures. The earliest method is called elemental powder sintering (EPS). In this technique, green compacts composed of mixed nickel and titanium elemental powders are sintered under argon at 900 ~ 1200 °C for different time. The sintering temperature and sintering time influence the phase composition, pore size, porosity, and mechanical properties. Although this process is simple, the maximum porosity achieved by this method is below 45 vol%, and the product does not possess large recovery strain[14, 15]. In addition, the pore size cannot be controlled precisely and the sintered products have more secondary phases which impair the SE and mechanical properties of these porous SMAs. Another frequently used method is the traditional hot isostatic pressing (HIP) process. Lagoudas et al. have prepared porous NiTi by this method.[16] However, the compressive strength is still low possibly due to the irregular pore shape and pore distribution in the porous NiTi. Greiner et al. have fabricated porous NiTi alloys by mixing pre-alloyed martensitic NiTi powders and small amount of elemental Ni powders using HIP. Although these porous products show excellent SE and higher compressive strength, the reported maximum porosity of the porous NiTi prepared by this method is lower than 20% and the size of most pores obtained by this process is only 10-50 μm.[17] Combustion synthesis (CS) is also usually employed to fabricate porous NiTi SMAs, and this technique fully utilizes the released energy from exothermic reactions. The CS process has two modes including self-propagating high-temperature synthesis (SHS) and thermal explosion. In the SHS process, the green compacts are put into a furnace filled with a protective gas and ignited after preheating at various temperatures. In this way, exothermic reactions (1), (2) and (3) take place along the green bar and the porosity of the porous NiTi can be controlled easily by changing the parameters. Using this method, anisotropic porous structures can be easily achieved and it can be attributed to the convective flow of the liquid and argon during combustion. Li et al. used this method to synthesize porous NiTi SMAs with larger porosity and linear channels.[18] In comparison with EPS and HIP, higher porosity and big pore size in the range of 50-70 vol% and 200–600 μm, respectively can be achieved by this method.[3,4] In the case of thermal explosion, there is no preheating process, and exothermic reactions occur instantaneous in the whole compacts at the same time when heated to a critical temperature. In comparison with SHS, this mode saves time, but it cannot be controlled easily. The porous NiTi samples produced by the two modes of CS exhibit lower compressive strength.[4,19,20]

New PM processes such as spark plasma sintering (SPS) and metal injection molding (MIM) have recently been developed to fabricate porous NiTi SMAs. SPS is mainly characterized by a spark plasma created by a pulsed direct current during heat treatment of the powders in the graphite die. The high-energy plasma is generated between the gap of two electrodes in the electric discharge machine[21]. The reported maximum porosity of porous NiTi SMAs prepared by SPS is lower than 30 vol% and the corresponding sample exhibits lower compressive strength[22]. The powder injection molding process which uses ceramic or metal powders is commonly termed metal injection molding (MIM). The MIM process starts with the mixing of metal powders with one or more polymers to produce a feedstock, followed by molding of the homogenized feedstock into shaped parts, removal of the polymer, and finally sintering. Bram et al. fabricated NiTi shape memory alloy parts using MIM, and they used pre-alloyed NiTi powders as the starting materials.[23] Recently, Hu et al. fabricated porous NiTi scaffolds using MIM in conjunction with SHS. The starting materials were Ni and Ti elemental powders and the scaffold produced by their process exhibited a total porosity of up to 75% with a maximum pore size of 200 μm.[24]

EXPERIMENTAL PROCEDURES

Fabrication methods of porous NiTi scaffold

Titanium and nickel powders with an average particle size of about 75 μm (purity > 99.5%) were put into a polymer can in equiatomic proportion together with some stainless steel balls. The powder to ball ratio was 1:2 by weight. The materials were mixed in a horizontal universal ball mill at a low speed of 100 rpm for 12 hours to produce a homogeneous mixture of the elemental powders and to minimize pre-alloying and oxidation of nickel and titanium powders during mixing. The mixture was combined with space holders (NH_4HCO_3 powders, purity > 99.5%) with a ratio of 3:10 (NH_4HCO_3 powders: Mixing powders) by weight. The final mixtures were pressed into green compacts in a steel mold with a diameter of 16 mm using a hydraulic machine at a cold compaction pressure of 200 MPa. Before sintering, the green compacts were pre-heated to remove these space holders. The pre-treated green compacts were subsequently put into capsule-free stainless steel canisters and sintered in the HIP unit to obtain porous NiTi. Details of the capsule-free hot isostatic pressing process can be found in our previous publication.[25] Samples 5 mm in diameter and 2 mm thick were cut from the sintered porous NiTi bars, mechanically ground, cleaned ultrasonically in acetone, and air-dried at room temperature.

Surface nanorization of porous NiTi scaffold

Surface nanorization of porous NiTi was achieved using a facile hydrothermal reaction. The 3D porous NiTi plates were put in a Teflon-lined autoclave together with 40 ml of 10 M NaOH. The autoclave was heated to 60 °C for different durations. After the hydrothermal treatment, the porous NiTi plates were fully washed by deionized water and dried in an oven at 60 °C for 12 hours. Phase identification was carried out by X-ray diffractometer (Siemens 500) using Cu K_α radiation at 40 kV and 30 mA. The film was analyzed using a scanning mode at 2° grazing incidence. The surface morphology of the samples was examined using field-emission scanning electron microscopy (FESEM, JEOL JSM-6335F). The microstructure and elemental composition of the nanobelts/nanowires were evaluated by transmission electron microscopy (TEM, Philips CM20) equipped with energy dispersive X-ray spectrometer (EDX). Fourier transform infrared (FTIR) spectra on the Perkin Elmer 16PC. The hydrophilicity of the plates was determined by measuring the surface contact angles on a Rame-Hart Mode 200, USA.

RESULTS AND DISCUSSION

Characterization of macro-porous structures

Capsule-free hot isostatic pressing (CF-HIP) is quite different from traditional HIP because the sintered compacts are in contact with the protective gas under high pressure. This process generally induces the porous structure in the sintered materials. Using this method, Yuan et al. have fabricated porous NiTi SMAs with a porosity of 40% with pore size ranging from 50 to 200 μm.[26] Although the porous materials have good mechanical properties, it is relatively difficult to adjust the pore size, distribution, and porosity by CF-HIP. We have fabricated porous NiTi SMAs with adjustable porous structures and good mechanical properties using CF-HIP with NH_4HCO_3 as the space holder. Figure 1 shows the typical porous structure of this scaffold. The pore size which is in the range of 50 ~ 800 μm is suitable for tissue ingrowth.[27]

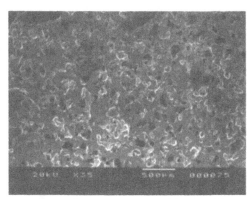

Figure 1. Typical porous structure of porous NiTi scaffold fabricated by CF-HIP with NH_4HCO_3 as the space holder.

The pore size, shape, and porosity can be controlled by varying the space holder content.[25] We have achieved general porosity and open porosity of about 56% and 70%, respectively as indicated by the interconnection of most pores.

Characterization of surface nano-structures

Preparation methods such as layer-by-layer processing,[28] pH-induced self-assembly,[29] colloidal self-assembly,[30] electron beam lithography (EBL),[31] and interference lithography (IL)[9] have been reported. However, these nano patterning and assembly techniques are not suitable for the large scale nano-patterning of porous NiTi scaffolds with complex surface topography.

It is obvious that potential techniques suitable for surface nanorization of the scaffolds should have a non-line-of-sight nature. Furthermore, the key step should be carried out in solutions because solutions in order to not only reach the entire exposed surface but also provide a suitable environment for the formation of nano materials on the exposed surface with no impairment to the SE and SME properties of porous NiTi SMA. In view of these factors, a low temperature hydrothermal method, i.e. alkaline solution treatment, is appropriate for surface nanorization of porous NiTi SMA scaffolds.

By means of this method, we have fabricated nanowires and nanobelts on the surface of porous NiTi scaffolds.[12] In the early stage, a nanoskeleton layer forms on the surface as indicated by Figure 2(a). As the reaction proceeds, a large number of one-dimensional (1D) nanowires and nanobelts form on this nanoskeleton layer as shown in Figure 2(b). These nanowires and nanobelts continue grow with the extension of reaction time [Figure 2(c)].

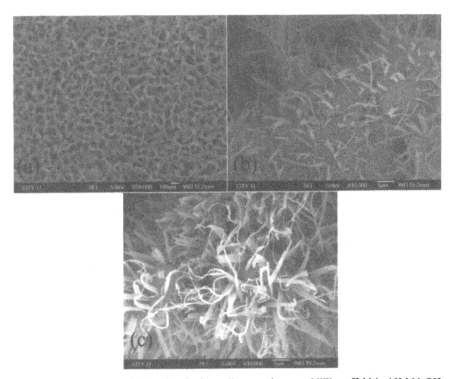

Figure 2 Surface morphologies of hydrothermally treated porous NiTi scaffold in 10M NaOH solution at 60°C with different duration time, (a) 4days, (b) 10days, and (c) 12days.

As shown in Figure 3, since the main peaks in the XRD pattern can be indexed to $H_2Ti_2O_5 \cdot H_2O$ with an orthorhombic lattice.

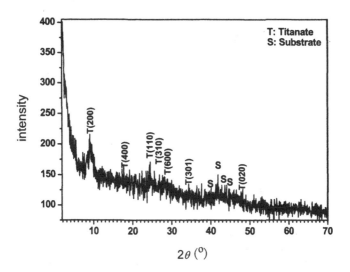

Figure 3. XRD patterns for hydrothermally treated 3D porous NiTi scaffold at 60 °C for 10days.

There is no massive ion exchange between sodium and hydrogen in this case because our experimental system cannot supply a large amount of H^+. EDS also indicates the existence of sodium (not shown here). Therefore, it is believed that these nanowires and nanobelts are predominantly composed of $Na_2Ti_2O_5 \cdot H_2O$. The selected area electron diffraction (SAED) pattern also confirms that these nano materials formed on the surface consists of a main phase of $Na_2Ti_2O_5 \cdot H_2O$ with only a trace of other phases as shown in the insert in Figure 4. The length of these nanowires/nanobelts can reach over 10 micrometers. The formation mechanism and crystal structure of these nano-titanates are still controversial. Some researchers ascribe the formation of the titanates to the direct reaction between metallic Ti and sodium hydride solution.[32] However, more researchers believe that the reaction is induced the formation of nanotitanates.[33] We have observed a native layer of titanium dioxide on the exposed area due to the higher temperature in PM in spite of the 99.995% protective argon atmosphere:

$$2TiO_2 + 2NaOH \rightarrow Na_2Ti_2O_5 \cdot H_2O \tag{4}$$

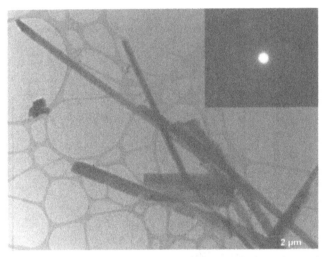

Figure 4. TEM image of scratched nanowires/nanobelts from hydrothermal treated porous NiTi scaffold. The insert image is the SAED pattern acquired from the nanobelt.

It can be observed from Figures 2(b) and (c) that despite the complex surface topography, nanowires and nanobelts can grow on almost the entire exposed surface, and the hierarchical scaffold here has no distinct interface between the substrate and nanostructured layer. Our previous XPS results reveal that the modified surface layer after the chemical treatment exhibited a graded structure.[34] In comparison with the surface self-assembled nanofiber matrix,[35] this graded structure can provide better bonding strength between the surface layer and substrate. This is very important to the long term fixation of this scaffold in the human body as well as minimization of the risk from debris released from the nanostructured surface. The relationship between the bonding strength and interfacial structure is being investigated and we will report our new findings in due course.

Our previous work indicates that the nanostructured surface layer can exhibit good hydrophilicity due to the unique structure of $Na_2Ti_2O_5 \cdot H_2O$.[12] FTIR examination shows the existence of a large amount of hydroxyl groups in these nanotitanates,[12] ensuring good hydrophilicity on the surface layer. In addition, the surface 1D nanostructure on the modified porous NiTi resembles the basic organization of human bone much better on the nano scale. Recent research has shown that these nanostructured scaffolds have larger surface areas to adsorb proteins. There are more binding sites to cell membrane receptors, and thus the adsorbed proteins may also change conformation exposing additional cryptic binding sites.[36] Our cell culture test has also indicated that more cells can adhere on the nanostructured surface compared to the untreated sample,[12] and the results are in good agreement with the theory proposed by Stevens and George.[36]

CONCLUSIONS

In order to improve the biocompatibility of porous NiTi SMA scaffolds in bone implants, a hierarchical structure resembling the organization of natural bone on both the macro scale and nano scale is necessary. Using capsule-free hot isostatic pressing methods with a space holder, we have fabricated porous NiTi SMA scaffolds that possess a macro-porous structure similar to that in human bones. Our recent work reveals that a concentrated alkaline solution treatment at low temperature is a facile and very practical method to modify the entire surface of porous NiTi scaffolds with complex topography. This low temperature technique can induce *in situ* growth of one-dimensional nanowires and nanobelts composed of $Na_2Ti_2O_5·H_2O$ as well as the underneath nanoskeleton layer on the exposed surface of porous NiTi scaffolds. These nanotitanates resemble the lowest organization of human bones on the nano scale. Our investigation also demonstrates that the hydrothermal duration significantly influences the surface morphology of the nano layer. The nanoskeleton layer forms in the early stage and 1D nanowires and nanobelts subsequently nucleate and grow on this layer. This nano-structured layer has good hydrophilicity favoring the cell adhesion and proliferation.

ACKNOWLEDGEMENTS

The work was financially supported by Hong Kong Research Grants Council (RGC) General Research Funds (GRF) No. CityU 112306 and 112307.

REFERENCES

1 J.A. Helsen, H.Jurgen Breme, *Metals as Biomaterials,* 1st ed. (John Wiley & Sons Publisher, New York, 1998) p.73
2 J.S. Kim, J.H. Kang, S.B. Kang, K.S. Yoon and Y.S. Kwon. Adv. Eng. Mater. 6, 403 (2004).
3 B.Y. Li, L.J. Rong, Y.Y. Li and V.E.Gjunter. Metall. Mater. Trans. A-Phys. Metall. Mater. Sci. 31, 1867 (2000).
4 C.L. Chu, C.Y. Chung, P.H. Lin and S.D. Wang. Mater. Sci. Eng. A-Struct. Mater. Prop. Microstruct. Process. 366, 114 (2004).
5 G. Balasundaram and T.J. Webster. J. Mater. Chem. 16, 3737 (2006).
6 J.Y. Lim and H.J. Donahue. Tissue Eng. 13, 1879 (2007).
7 M. Goldberg, R. Langer and X.Q. Jia. J. Biomater. Sci.-Polym. Ed. 18, 241 (2007).
8 K.M. Woo, J.H. Jun, V.J. Chen, J.Y. Seo, J.H. Baek, H.M. Ryoo, G.S. Kim, M.J. Somerman and P.X. Ma. Biomaterials 28, 335 (2007).
9 M.J. Dalby, N. Gadegaard, R. Tare, A. Andar, M.O. Riehle, P. Herzyk, C.D.W. Wilkinson and R.O.C. Oreffo. Nat. Mater. 6, 997 (2007).
10 M.J. Dalby, N. Gadegaard, P. Herzyk, H. Agheli, D.S. Sutherland and C.D.W. Wilkinson. Biomaterials 28, 1761 (2007).
11 K.A. Hing. Philos. Trans. R. Soc. A-Math. Phys. Eng. Sci. 362, 2821 (2004).

12 S.L. Wu, X.M. Liu, T. Hu, P.K. Chu, J.P.Y. Ho, Y.L. Chan, K.W.K. Yeung, C.L. Chu, T.F. Hung, K.F. Huo, C.Y. Chung, W.W. Lu, K.M.C. Cheung and K.D.K. Luk. Nano Lett. 8, 3803 (2008).

13 T.B. Massalski, H. Okamoto, P.R. Subramanian and L. Kacprzak. *Binary Alloy Phase Diagrams.* (Materials Park, Ohio: ASM International, 1990) p. 2874.

14 B.Y. Li, L.J. Rong and Y.Y. Li. Intermetallics 8, 643 (2000).

15 S.L. Zhu, X.J. Yang, F. Hu, S.H. Deng and Z.D. Cui. Mater. Lett. 58, 2369 (2004).

16 D.C. Lagoudas and E.L.Vandygriff. J. Intell. Mater. Syst. Struct. 13, 837 (2002).

17 C. Greiner, S.M. Oppenheimer and D.C. Dunand. Acta Biomater. 1, 705 (2005).

18 B.Y. Li, L.J. Rong, Y.Y. Li and V.E.Gjunter. Acta Mater. 48, 3895 (2000).

19 Y.H. Li, L.J. Rong and Y.Y. Li. J. Alloy. Compd. 345, 271 (2002).

20 A. Biswas. Acta Mater. 5, 1415 (2005).

21 O. Mamoru. Mater. Sci. Eng. A-Struct. Mater. Prop. Microstruct. Process. 287, 183 (2000).

22 Y. Zhao, M. Taya, Y.S. Kang and A. Kawasaki. Acta Mater. 53, 337 (2005).

23 M. Bram, A. Ahmad-Khanlou, A. Heckmann, B. Fuchs, H.P. Buchkremer and D. Stover. Mater. Sci. Eng. A-Struct. Mater. Prop. Microstruct. Process. 337, 254 (2002).

24 G.X. Hu, L.X. Zhang, Y.L. Fan and Y.H. Li. J. Mater. Process. Technol. 206, 395 (2008).

25 S.L. Wu, C.Y. Chung, X.M. Liu, P.K. Chu, J.P.Y. Ho, C.L. Chu, W.W. Lu, K.M.C. Cheung and K.D.K. Luk. Acta Mater. 55, 3437 (2007).

26 B. Yuan, C.Y. Chung and M. Zhu. Mater. Sci. Eng. A-Struct. Mater. Prop. Microstruct. Process. 382, 181 (2004).

27 D. Tadic, F. Beckmann, T. Donath and M. Epple. Materialwiss. Werkstofftech. 35, 240 (2004).

28 E. Jan and N.A. Kotov. Nano Lett. 7, 1123 (2007).

29 J.D. Hartgerink, E. Beniash and S.I. Stupp. Science 294, 1684 (2001).

30 C. Huwiler, T.P. Kunzler, M. Textor, J. Voros and N.D. Spencer. Langmuir 23, 5929 (2007).

31 J.H. Jang, C.K. Ullal, T. Gorishnyy, V.V. Tsukruk and E.L. Thomas. Nano lett. 6, 740 (2006).

32 X. Peng and A Chen. Adv. Funct. Mater. 16, 1355 (2006).

33 X.M. Sun, X. Chen and Y.D. Li. Inorg. Chem. 41, 4996 (2002).

34 S.L. Wu, X.M. Liu, Y.L. Chan, P.K. Chu, C.Y. Chung, C.L. Chu, W.W. Lu, K.M.C. Cheung and K.D.K. Luk. J. Biomed. Mater. Res. Part A, DOI, 10.1002/jbm.a.32008.

35 T.D. Sargeant, M.O. Guler, S.M. Oppenheimer, A. Mata, R.L. Satcher, D.C. Dunand and S.I. Stupp. Biomaterials 29, 161 (2008).

36 M.M. Stevens and J.H. George. Science 310, 1135 (2005).

Iñigo Braceras[1,2], Iñaki Alava[1,3], Roberto Muñoz[1,3], and Miguel Angel De Maeztu[4]
[1] Inasmet-Tecnalia, Mikeletegi Pasealekua 2, 20009 Donostia-San Sebastian, Spain
[2] Lifenova Biomedical SA, Mikeletegi Pasealekua 2, 20009 Donostia-San Sebastian, Spain
[3] CIBER-BBN, Spain
[4] Private Practice, Pº San Francisco, 43-A1, 20400 Tolosa, Spain

ABSTRACT

A key process in a successful treatment of patients with a great variety of musculoskeletal implants requires a fast, reliable and consistent osseointegration. Among the parameters that affect this process, it is widely admitted that implant surface topography, surface energy and composition play an important role.

Different surface modification techniques to improve osseointegration have been proposed and tested to date, but most focus on microscale features, and few control surface modifications at nanoscale. On the other hand, ion implantation modifies the outermost surface properties in relation to the nanotopography, chemical and physical characteristics at nanoscale. The meta-stable surface that results from the treatment, affects the adsorption of bio-molecules in the very first stages of the implant placement, and thus the signaling pathway that promotes the differentiation and apposition of osteoblast cells.

This study aimed at assessing the performance, in terms of osseointegration levels and speed, of ion implanted titanium made implants. The study included several in vitro and in vivo tests. The latter, comprised different insertion periods and both experimental and commercial implants as comparative surfaces. The final stage of the study included clinical trials in human patients.

In each and every case, bone integration improvement of tested materials/implants was achieved for the CO ion implanted samples. Furthermore, contact osteogenesis was observed in the ion implanted samples, unlike the Ti control samples, where only distance osteogenesis occurred, being this potentially one of the reasons for their faster healing and osseointegration process.

Finally, the use of ion implantation as a surface modification tool that allows for evaluating the effects of nanotopography and composition changes independently is presented.

INTRODUCTION

Load bearing implants are widely used today in the clinical field, with titanium and its alloys the material of choice in bone engaging components. This is so because among available metallic alloys, titanium offers the best combination of bulk mechanical properties and good surface properties, and it is actually well established that osseointegration occurs on Ti surfaces. Nevertheless, shortcomings in terms of patient treatment times, i.e. time required for osseointegration to happen and thus allow a safe loading of the implant, and failure rates have led to important research activity on the surface properties to promote a faster and more complete osseointegration and lower wear rates. Thus surface chemistry, physical properties and topography, which are commonly accepted to regulate the tissue response [1], have been the focus of intense research.

Nowadays, macro & micron scale surface modifications are regularly applied on a commercial basis on Ti and its alloys to promote osseointegration: sandblasting, acid etching based solutions and its combinations, thermal spraying (either atmospheric plasma spraying of hydroxyapatite or titanium plasma spraying). One accepted route in attempting to enhance bone differentiation and promote fast and direct bone growth on implants is based on studying the underlying reactions at nanoscale [2,3,4,5,6,7]. Among different techniques under study, ion implantation based treatments have also been explored in the last years. Some of the difficulties many of these techniques find when trying to explain the effect of nanoscale modifications on osseointegration lays in discerning between the separate effects of topographical, chemical and physical surface modifications [8], and in the study of the events occurring at the material and living tissue interface at the nanoscale [9].

In the present study, the effects of CO ion implantation have been studied. Besides the surface analysis of titanium subjected to this treatment, the evaluation of its effects on osseointegration is presented at several stages, from early in vitro studies to final clinical trials in human patients.

Furthermore, in order to understand what role features such as nanotopography or surface chemistry play, the use of ion implantation to produce chemically similar surfaces with different nanotopographies is described.

EXPERIMENT

TiGr5 and TiGr4 samples were subjected to two set of ion implantation treatments respectively, which were performed in a Danfysik 1090 High Current ion implanter. The first set of ion implantation treatments consisted on CO ion implantation of polished TiGr5 discs for the surface analysis and in vitro tests and of commercial TiGr5 made dental implants for the in vivo tests. The second set consisted on Ne, Ar, Kr, Xe ion implantation at different energies (40keV and 80keV) and fluences ($1x10^{17}$ and $2x10^{17}$ ion/cm^2) of polished TiGr4 discs.

Surface analyses were performed by X-Ray Photoelectron Spectroscopy (XPS) in a Microlab MKII VG spectrometer, by Atomic Force Microscopy (AFM) in a MultiMode™ Digital Instruments equipment with a "J" type scanner in tapping mode, and by contact angle, threefold per sample in different areas, by computerized image analysis (DIGIDROP from GBX Instruments) using deionised water with a constant and controlled drop size.

In vitro evaluation of the first set of treatments was performed by cell apoptosis tests of human bone cells, after incubation for 24 and 72 h at 35 °C and 5% CO_2 on treated and untreated control samples. The apoptosis state of the cells was tested using a flow cytometry Kit (BD™ Cytometric Bead Array. Human Apoptosis Kit), which detects the concentrations of Poly (ADP-ribose) polymerase or PARP, Bcl-2 and Caspase-3 molecules. The expression of alkaline phosphatase (ALP) was also studied with human bone cells, which were incubated for 3, 8 and 16 days at 35 °C and 5% CO_2; the activity was determined measuring the final fluorescence emitted as a result of an enzyme based reaction (4-Methylumbelliferyl phosphate + ALP).

In vivo tests were performed on the mandible of Beagle dogs, where premolars had been removed three months before and where implants stayed for 3 and 6 months. The surgical protocol, duly approved by the ethical committee, has already been described elsewhere [10]. The study compared the histomorphometric results in terms of osseointegration for implants surface treated using CO ion implantation with three other surface treatments, acid-etched, sandblasted/acid-etched and anodic oxidation, as well as with machine-turned titanium implants as a control group.

The clinical trial (Ref. 277/06/EC) consisted on a multicenter study, which had been previously approved by the corresponding ethical committees and legal authorities (AGEMPS, Spanish Agency for Drugs and Medical Devices), and was performed with mini implants, where half the longitudinal surface had been subjected to the CO ion implantation treatment, while the other half remained untreated, in the as-machined condition.

In both the in vivo tests and clinical trials, the retrieved samples were analyzed to quantify the bone implant contact percentage (BIC%) by two different techniques: first by Environmental Scanning Electron Microscopy (ESEM) (JEOL JSM-5910LV, Akishima City, Tokyo, Japan) (figure 1) and Energy Dispersive Spectrometry (EDS) (INCA 300), and then, after preparation of thin samples by embedding in resin, sectioning, fine polishing into 50-μm thick samples, and finally, toluidine blue staining, by conventional light transmission microscopy using a digital microphotography system (Nikon Kodak® Ltd, Rochester, NY, USA), the Adobe Photoshop 7® program (Adobe System Inc., San Jose, CA, USA) and the Omninet® image analysis system (Buehler Ltd, Lak Fluff, IL, USA) (figure 2).

Figure 1. ESEM image of a histological preparation of an ion implanted dental implant for evaluation of the osteointegration, x35.

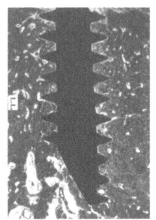

Figure 2. Image of a thin histological preparation of an ion implanted dental implant for evaluation of the osteointegration, x12.

DISCUSSION

The results and discussion concerning the CO ion implantation treatment are presented first. Next, the surface analysis results related to the second set of ion implantation treatments is presented together with the conclusions.

CO ion implantation and osseointegration

Surface analysis of CO ion implanted TiGr4 samples evaluated the surface roughness, surface chemistry and wettability. The roughness, as measured by AFM, was observed to increase to Ra 1.9 to 2.6 nm depending on the treatment conditions (vs. Ra 0.5 nm for control samples). Contact angle did also increase up to 93.8°±2.7 (vs. 72.9°±7.9 for control samples), thereby making the surface mildly more hydrophobic [11]. XPS analysis showed that the profile of the ion implanted region went down to a depth of around 60nm, with a variety of C rich compounds.

In vitro tests (figure 3) showed after 24 and 72 h of incubation that all apoptosis signals, PARP, Bcl-2 and Caspase-3 molecules, were lower on CO ion implanted samples relative to the control sample. For PARP, values were 1136.0±205.4pg/ml and 782.3±180.8pg/ml at 24hours, and 393.6±44.8pg/ml and 183.3±133.7pg/ml at 72 hours for the control and CO ion implanted samples respectively. This lower apoptosis signal is a predictor of a better evolution of the osteoblasts on the treated surface.

On the other hand, alkaline phosphatase activity was higher on the CO ion implanted samples, with a strong significant difference in the early stages (first test period of 3 days):

1124.98±218.07 µmol/mg protein/min vs. 685.14±208.48 µmol/mg protein/min on untreated control samples. Alkaline phosphatase activity increases as pre-osteoblasts differentiate into osteoblasts. Therefore, the expression of this enzyme is frequently used as a marker of osteoblast differentiation [12]. Thus this is a clear indication that since the ion implanted surface increases ALP activity, osteoblast differentiation and accelerated osseointegration will occur.

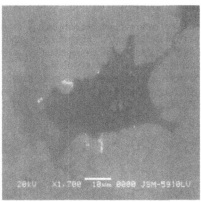

Figure 3. ESEM image of an osteoblast on a CO ion implanted surface, x1,700.

In the case of the in vivo tests with dental implants, the histomorphometric analysis by ESEM of the samples retrieved showed that the percentage of bone contact for each group at 3 and 6 months respectively was: ion implantation 61% and 62%; acid-etched 48% and 45%; sandblasted/acid-etched 46% and 52%; anodic oxidation 55% and 46%; control 33% and 49% [9]. All implants with treated surfaces exhibited thus a higher %BIC from an early stage vs. the control group. For the control group %BIC values were significantly higher 6 months after placement than at month 3, unlike in the case of all the surface-treated implants, where osseointegration values were similar at months 3 and 6. The greater osseointegration observed at early stages on implants with treated surfaces is in accordance with the results obtained elsewhere [13,14] . Among the different surface treatments included in this study, implants treated with CO ion implantation showed a higher percentage of BIC at 3 and 6 months after placement than the average of all the other treatments when grouped. Furthermore, these differences were statistically significant when comparing ion implantation with all the other treatments grouped together and with the control group at 3 and 6 months. It is also worth mentioning that no statistical differences were observed between the three commercial micro-scale treatments and the untreated control samples, or between themselves, at either 3 or 6 months. Therefore, according to both evaluation methods, CO ion implantation improved osseointegration in comparison to the classical micro-scale surface treatment methods.

Clinical trials with the CO ion implantation treatment have recently been successfully completed, with 19 patients involved in the study. 22 mini-implants retrieved from the patients were analyzed and the BIC% evaluated. The results confirmed again that CO ion implanted dental implants presented a larger osseointegration than the as machined samples. Moreover, no negative reaction was observed in the patients, thus concluding that CO ion implanted Ti surfaces suppose no risk whatsoever to human patients.

Nanotopography & ion implantation

Ion implantation into Ti alloys might not only change the surface chemistry, but does also change the topography at nanoscale. Both are recognized as important parameters affecting the osseointegration of Ti. With a second set of ion implantation treatments, the independent effect of nanotopography on osseointegration is being evaluated.

For such an evaluation, ion implantation of Ne, Ar, Kr, Xe at different energies and fluences have been applied on polished TiGr4, producing remarkably different surface topographies (e.g. see Fig. 4).

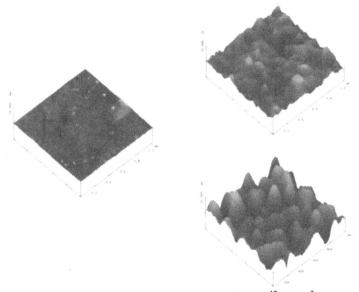

Figure 4. AFM images (1x1 μm) of polished (left), Ar (40keV; $2x10^{17}$ ion/cm^2; upper right) and Xe ion implanted (40keV; $2x10^{17}$ ion/cm^2; bottom right) TiGr4 samples.

The nanoroughness generated by the ion implantation ranged from Ra 0.56 nm to Ra 7.27 nm (vs. Ra 0.30 nm for control samples). The effect of these ion implantation treatments on wettability was relatively mild. Contact angle measurements ranged from 78.6°±1.6 to 88.8°±2.1 (vs. 82.6°±5.2 for control samples). Furthermore, it did not significantly change in the case of larger ions (Kr and Xe).

CONCLUSIONS

This study shows that surface treated titanium does achieve a faster osseointegration. Nonetheless, no statistically significant differences have been found among the micro-scale surface treatments studied. On the other hand, CO ion implantation, which produces surface modifications at the nanoscale, showed a faster and larger osseointegration, being this supported by extensive testing that encompasses from in vitro lab tests to clinical trials.

Additionally, ion implantation is deemed to be a useful tool in producing a range of different nanoscale topographies, which can be valuable in differentiating the effect of topography from chemical modifications on osseointegration.

ACKNOWLEDGMENTS

Research work was partially supported by the IN-2005/0000018 grant and ETORTEK BIOSUPERFICIES, Basque Government IE07-201 research program.

REFERENCES

1. T. Albrektsson, P.I. Brånemark, H.-A. Hansson and L. Lindstrom, Acta Orthop. Scand. 52, 155 (1981).
2. G. Mendonça, D.B.S. Mendonça, F.J.L. Aragão and L.F. Cooper, Biomaterials 29(28), 3822-3835 (2008).
3. M.J. Dalby, D. McCloy, M. Robertson, C. D.W. Wilkinson and R.O.C. Oreffo, Biomaterials, 27(8), 1306-1315 (2006).
4. G. Zhao, A.L. Raines, M. Wieland, Z. Schwartz and B.D. Boyan, Biomaterials, 28(18), 2821-2829 (2007).
5. D. Khang, J. Lu, C. Yao, K.M. Haberstroh and T.J. Webster, Biomaterials, 29(8), 970-983 (2008).
6. L. Zhang, T.J. Webster, Nano Today, 4(1), 66-80 (2009).
7. T.J. Webster and J.U. Ejiofor, Biomaterials, 25(19), 4731-4739 (2004).
8. R.G. Flemming, C. J. Murphy, G.A. Abrams, S.L. Goodman and P. F. Nealey, Biomaterials 20(6), 573-588 (1999).
9. L. Marcotte and M. Tabrizian, IRBM 29(2-3), 77-88 (2008).
10. M.A. De Maeztu, I. Braceras, J.I. Alava and C. Gay-Escoda, Int J Oral Maxillofac Surg 37(5), 441-447 (2008).
11. I. Braceras, J.I. Alava, L. Goikoetxea, M.A. de Maeztu, J.I. Onate, Surf. Coat. Technol. 201(19-20), 8091-8098 (2007).
12. J.E. Aubin, F. Liu, L. Malaval and A.K. Gupta, Bone 17, 77S (1995).
13. G. Orsini, B. Assenza, A. Scarano, M. Piatteli and A. Piatteli, Int J Oral Maxillofac Implants 15, 779–784 (2000).
14. D.L. Cochran, D. Buser, C.M. Ten Bruggenkate, D. Weingart, T.M. Taylor, J.P. Bernard, F. Peters and J.P. Simpson, Clin Oral Implants Res 13, 144–153 (2002).

Mater. Res. Soc. Symp. Proc. Vol. 1181 © 2009 Materials Research Society   1181-DD09-01

Low-Energy Ion Beam Biotechnology at Chiang Mai University

L.D. Yu[1,2]* and S. Anuntalabhochai[3]
[1] Plasma and Beam Physics Research Facility (PBP), Department of Physics, Faculty of Science, Chiang Mai University, Chiang Mai 50200, Thailand
[2] Thailand Center of Excellence in Physics, Commission on Higher Education, 328 Si Ayutthaya Road, Bangkok 10400, Thailand
[3] Molecular Biology Laboratory, Department of Biology, Faculty of Science, Chiang Mai University, Chiang Mai 50200, Thailand
(*Corresponding author: Email: yuld@fnrf.science.cmu.ac.th)

ABSTRACT

MeV-ion beam has long been applied to research and applications in biology for many decades as highly energetic ions are undoubtedly able to interact directly with biomolecules to cause biological changes. However, low-energy ion beam at tens of keV and even lower has also been found to have significant biological effects on living materials. The finding has led to applications of ion-beam induced mutation and gene transfer. From the theoretical point of view, the low-energy ion beam effects on biology are difficult to understand since the ion range is so short that the ions can hardly directly interact with the key biological molecules for the changes. This talk introduces interesting aspects of low-energy ion beam biology, including basis of ion beam biotechnology and recent developments achieved in Chiang Mai University in relevant applications such as mutation and gene transfer and investigations on mechanisms involved in the low-energy ion interaction with biological matter such as eV-keV ion beam bombardments of naked DNA and the cell envelopes.

INTRODUCTION

Plant seeds were carried by spacecrafts and sent to the space for mutation purpose because high-energy cosmic particles might irradiate and penetrate the seeds to induce mutation. However, scientists have found that they can achieve the same effect on the earth ground using low-energy particle bombardment of the seeds in a much cheaper, easier and more effective way. That is the initiation of the low-energy ion beam technology application in biology. Low-energy ion beam biotechnology (IBBT) is such a technique that uses energetic ion beams (a few tens of keV in energy being enough), generated and transported by an ion accelerator, to bombard biological organisms in vacuum to induce mutation breeding and gene transfer for increasing applications of the biological substance as well as to detect and analyze characteristics of biological organisms [1]. Ion beam interaction with biological living organisms is so different from that with normal solid materials. Biological organisms are living, and ion beam treatment should not kill them. Fresh cells contain a large amount of water, which essentially evaporates in vacuum, and the evaporation causes differences in the target status from that in normal atmosphere. Biological material structures are highly porous and inhomogeneous, and ions penetrate and sputter abnormally more than for normal condensed materials. The functioning structures of organisms are very complicated and different ion-beam treated locations have different responses, and hence in order to get a certain response, ion beam should be precisely controlled to target the location. Biological organisms are extremely sensitive to ion irradiation, will actively respond to the irradiation and thus highly produce secondary effects, which can greatly affect consequences of ion beam bombardment. Organisms have recovery ability, and ion beam radiation damage may be repaired and thus ion beam effects may disappear in a certain time period. Different parts of an organism may have communications and an ion-beam treated location may produce unexpected effects. This is a highly interdisciplinary field of physics, biology, agriculture and medical science. It is an extension of ion beam modification of conventional solid materials and an extension of high-energy radiobiology. This review talk introduces interesting aspects of low-energy ion beam biology, including basis of ion beam biotechnology and recent developments achieved in Chiang Mai University (CMU) in relevant applications such as mutation and gene transfer and investigations on mechanisms involved in the low-energy ion interaction with biological matter.

IBBT FACILITIES AT CMU

A bioengineering-specialized ion beam line has been self-developed, which possesses special features [2]: vertical main beam line, low-energy (30 keV) ion beams, double swerve of the beam, a fast pumped target chamber, and an *in-situ* atomic force microscope (AFM) system [3] chamber, as shown in Fig.1. The whole beam line is situated in a bioclean environment, occupying two stories. The quality of the ion beam has been studied. It has proved that this beam line has significantly contributed to our research work on low-energy ion beam biotechnology. Another facility is a non-mass-analysis ion implanter [4], which was originally developed for industrial applications and later modified for bioengineering purpose as well. This installation is featured with high current, low-to-medium energy applicability, large implantation area, and versatile sample holder and stage with water cooling. This facility is particularly used for ion beam mutation induction.

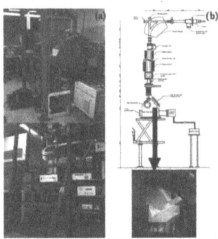

Figure 1. The bioengineering ion beam line at CMU. (a) Photograph. The upper one is the part upstairs and the lower one is the part downstairs. (b) Schematic diagram with a photo of the *in-situ* AFM sitting inside the chamber.

ION BEAM INDUCTION OF GENE TRANSFER

Introduction

Using ion beam bombardment is a physical method to induce DNA transfer in cells. In the method, the ion beam parameters such as ion species, energy and fluence are precisely controlled so that the ions can only bombard the cell envelope to crate radiation-damage-induced special structures which are expected to act pathways for exogenous macromolecules to pass through. The range and extent of the radiation damage for the target material of the cell envelope depend on the ion beam parameters as mentioned above. Therefore, the ion beam parameters should be carefully designed for treating different biological cell species which have different cell envelope thicknesses. The post biological treatment for DNA transfer is carried out immediately after ion beam bombardment normally in standard biological procedures.

DNA transfer

Using the critical ion beam conditions, we succeeded in DNA transfer in bacterial cells of *E. coli* [5,6]. Fig. 2 shows the introduction of pGEM-T easy and pGFP plasmids, containing *Lac z* and *GFP* genes respectively, into *E. coli* strain DH5α bombarded by Ar ions at an energy of 26 keV and a fluence of 2×10^{15} ions/cm^2. In the experiment, the bacterial cells were directly bombarded by the Ar-ion beam in vacuum, followed by a standard DNA transfer process in solution. We observed the plasmid DNA indeed transferred in the ion-bombarded bacterial cells but not in the unbombarded control. The blue and green colonies observed under UV at the bacteria incubated in medium are the indicators of the marker genes *Lac z* and *GFP*, respectively, and thus demonstrate that the DNA has been indeed transferred into the bacterial cells. The subsequently measured DNA molecular sizes further confirm that the transferred DNA is the original exogenous DNA. In another work, two marker genes, GFP and *lipoic acid synthetase,* were chosen to transfer into yeast (*Saccharomyces cerevisiae* strain W303C, *MATa ura3-52 his3-Δ200 lys2-801*) [7]. The yeast cells were bombarded by nitrogen ions at energy of 50 keV with fluences of 0.5, 1, and 2×10^{16} ions/cm^2, and subsequently the bombarded yeast cells were incubated with both plasmids, pYGFP and pYlip plasmids, carrying *GFP* and *lipoic acid synthetase* genes separately. The expression of *GFP* gene in yeast was observed by green yeast colonies under UV light (Fig. 3a), while the expression of *lipoic acid synthetase* gene was analyzed using SDS-PAGE method. A gene's product at 34 kDa was detected only in the bombarded yeast with the pYlip (Fig. 3b). The expression of both genes was induced by culturing the yeast cells in YPD (yeast potato dextrose) media supplemented with galactose for 10 hrs. These evidences demonstrated that nitrogen ions assisted gene transfer into the yeast cells.

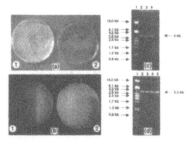

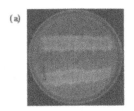
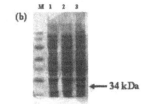

Figure 2. Expression of the ion-beam-induced gene transfer in the cells of bacteria *E. coli* [(a), (b)] and the corresponding molecular size measurement [(c), (d)] by electrophoresis. (a) and (c): transferred pGEM-T easy plasmid. (b) and (d): transferred pGFP plasmid. (1) *E. coli* without gene, and (2) with gene transferred. In (c) and (d), lane 1 shows the standard molecular size markers, and other lanes show the DNA molecular sizes measured by various restriction enzymes.

Figure 3. Demonstration of ion-beam-induced transfer of plasmid DNA, pGFP and pYlip containing *GFP* and *lipoic acid synthetase* genes, respectively, in yeast (*S. cerevisiae* strain W303C) bombarded by nitrogen ions at energy of 50 keV and fluences of 2×10^{16} ions/cm^2. (a) A green colony of yeast indicates an expression of *GFP* gene in the yeast cells. (b) SDS-PAGE analysis of *lipoic acid synthetase*; (1) *S. cerevisiae* wild-type, (2) transformed but non-induced, (3) transformed and induced recombinant *S. cerevisiae*, and (M) molecular mass marker. The arrow indicates the band (M~34 kDa) corresponding to recombinant lipoic acid synthetase. The gel was stained with Coomassie blue to visualize the protein bands.

Investigations on mechanism

Mechanisms involved in ion beam induction of gene transfer have been investigated. We found that subjected to ion bombardment, micro- and nano-craters were formed on the biological cell surface of either plants or bacteria [3,8-10], as shown in Fig. 4. These tiny craters might function as pathways for the exogenous macromolecule to transfer into cells. A physical model was set up for ion implantation of the cell wall materials to explain abnormally great ion range and sputtering in the cell wall [11]. The model demonstrated that with appropriately low ion energy and fluence, ions were able to penetrate through the cell envelope and possibly create the pathways.

For separately investigating ion bombardment effect on the cell envelope materials, we used chemically similar chitosan and cellulose membranes to substitute for the cell envelope subjected to ion bombardment. It was found that the electric impedance decreased and the capacitance increased and these electric property changes enhanced DNA transfer through the membrane [12]. Electron-neutralized ion beam was also applied to bombard the membranes to check charge effect. It was found that the neutralized beam reduced the impedance pronouncedly compared with the ion beam [13].

Figure 4. Low-energy ion bombardment induced formation of micro/nano-craters in the plant cell envelope. (a) SEM photograph of the surface of the unbombarded onion skin cells. (b) *Ex-situ* SEM photograph of the 30-keV Xe-ion bombarded onion skin cell. Bar scale: 2 μm.

ION BEAM INDUCTION OF MUTATIONS

Introduction

Induction of mutation by external energetic particles stems from the idea that the particles penetrate through the coat of the germinating parts of biological organisms such as the embryo of a seed or the buds and then interact with the genetic substance in the cells of the germinating parts. In normal operation conditions, biological organisms are put in vacuum environment and subjected to ion beam bombardment. Therefore, special measures should be taken. In the case of seeds, if the seed is big and its coat is thick, it is better to remove the seed coat covering the embryo part beforehand to directly expose the embryo part to the ion beam for high-efficiency mutation induction. In case the seed is too small, if the embryo location is known it is better to orient the embryo part towards the ion beam; if the embryo location is unknown, the seeds have to be positioned in random for ion bombardment. In the case of tissues such as buds, the non-critical parts of the tissue should be well wrapped to prevent the water-evaporation worsening the vacuum and to expose only the germinating part to the ion beam. Here we introduce some examples of our successes in ion-beam-induced mutations and relevant studies on mechanisms.

Mutation induction

Two varieties of local rice (*Oryza sativa indica*) [14], purple glutinous rice (Kum Doi Saket) and Thai jasmine rice (KDML 105) [15], were chosen for the mutation induction purpose. Thousands of seeds of the rice were implanted with nitrogen ions at energy of $30 - 60$ keV to fluences of an order of 10^{16} ions/cm^2 and mutants were selected from grown plants of the seeds. For the purple glutinous rice, seedlings with green pigment were observed in M1 generation while the wild type was purple. Seeds in the M1 generation were harvested and cultivated for M2 generations. Their phenotypes were divided into 3 groups: 1) the whole plant was still green, 2) only the stem was green while the leaf blade and sheath were purple, and 3) the whole plant was purple. It was also observed that the pigment of the seat coat and pericarp were not purple as was found in the wild type (Fig. 5). In a mutant, the starch content in the harvested seeds showed a blue-black color versus the normal color when stained with I_2 solution (Fig. 5). This indicated that low energy ion beam induced mutation of glutinous rice. Analysis of the DNA fingerprint revealed genotypic differences among rice samples in the M2 generation, indicating genetic modification occurring in their genome. For Thai jasmine rice, various mutants were obtained after screening. Fig. 6 shows one of the examples in the straw stem height change. DNA sequencing analysis obtained two DNA sequences. One belonged to a member of flavonoid

3'hydroxylase of *O. sativa japonica* with the highest identity of 60%. The flavanoid 3'hydroxylase is the enzyme that is involved in anthocyanin biosynthetic pathway. Anthocyanin pigments display color ranging from bright red/purple to blue. Color variation of the main red/purple in various tissues such as leaf sheath, collar, auricles, ligule, and pericarp, and light brown in starchy endosperm may be induced from mutation in genes controlling the purple/red to blue. Another was 61 % identity to cytochrome P450 of *O. sativa japonica*. It is a member of cytochrome P450, playing significant role in biological systems, such as hormonal regulation, phytoalexin synthesis, as well as flower petal pigment biosynthesis, probably resulting in the phenotypic variations.

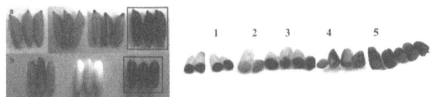

Figure 5. Ion-beam-induced mutations of purple rice. Left: phenotypes of M2 seeds. Pigment of seed coat (a) and pericarp (b) were not dark purple as in wild type (square box). Right: Starch content in purple rice seeds. After staining with aliquot of iodine solution the blue-dark color was observed in non glutinous rice seeds bombarded by nitrogen. Non glutinous Thai jasmine rice seed control (1), bombarded purple rice seeds with white pericarp from the same panicle (2 and 3), purple rice seed control (4) and non-glutinous black rice seeds (5).

Figure 6. Photograph of the ion-beam-induced mutants in M_5 generation, showing the straw stem height change.

Figure 7. One of the examples in ion-beam-induced color change in Gerbera flower petal. Note that the original color should be all in red.

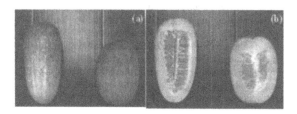 **Figure 8.** Effect of ion beam on reproductive organ of cucumber. (a) Size comparison of whole fruit of cucumber between control (left) and hybrid (right). (b) Fresh meat of control (left) and hybrid (right).

Seeds and buds of various local flowers such as petunia, rose, gerbera and chrysanthemum were bombarded by nitrogen ion beams at energies of 30 – 60 keV and fluences of $1 \times 10^{16} - 5 \times 10^{17}$ ions/cm^2 in vacuum [16]. After ion beam treatment, the seeds or buds were cultured to grow. After screening, various mutants were found with features of changes in the flower shape, color, petal shape and size, and petal color (Fig. 7). DNA analysis revealed different polymorphic bands in the mutants from in the control. Seeds of two vegetable species, cucumber (*Curcumis sativus*) and mung bean (*Vigna radiata*) were bombarded by N-ion beam at several tens of keV to fluences ranging from 10^{16} to 10^{18} ions/cm^2. For cucumber, about 3% of bombarded plants at 2×10^{17} ions/cm^2 showed their size of female reproductive organ (ovary) greater than control and some were transformed into male reproductive organ. Crossing pollination between the greater organ to male organ of wild type resulted in short of fruit size but normal fresh meat. (Fig. 8). For bean, the ion-beam-induced mutants showed less stem height, number of leaves, growth rate and dry mass than the control significantly.

Gene cloning

Anthracnose, caused by the fungus *Colletotrichum* sp., is one of the important diseases affecting flowers. The use of natural antagonists has recently been applied for biological control. In the study, the target material was *Bacillus* (*B.*) *licheniformis* (a kind of bacteria) isolated from hot springs in Chiang Mai, showing its activity to suppress conidia germination of the fungus and reduce symptom caused by the disease in flower plants. N-ions at energy of 28 keV were used to bombard the bacteria to a fluence range of $1 - 10 \times 10^{15}$ ions/cm^2. After ion bombardment, mutants were screened and one of the mutants with loosing its antagonistic property was obtained. For this mutant DNA fingerprint analysis was carried out. The additional band of the mutant in the DNA analysis was subcloned into pGEM-T easy vector and sequenced. Partial sequence analysis revealed that this fragment was a gene encoding enzyme lipase. Regarding to the lipase gene sequence, pair of specific primer was designed from the *B. licheniformis* database to amplify the entire sequence of the gene using its genomic as template. In order to determine the expression of the lipase gene in yeast, the entire lipase gene was subcloned into pYES2 and named pYES2-LicLip, then transferred into yeast cell via low energy ion beam. The result showed that only the yeast transformed with pYES2-Liclip induced colonies and *B. licheniformis* induced colonies presented antifungal activity to plant pathogen, collectotrichum musae [17].

Investigation on mechanisms

167

How low-energy ion beam can induce mutation is yet a puzzle. Theoretical calculations predict the ion range in the materials that cover DNA such as seed coat and cell envelope considerably shorter than the material thickness and thus it is impossible for the low-energy ions to act with DNA. However, the reality may be far deviated from the assumption. The plant seed coat may be significantly porous and hence the stopping power of the seed coat to the incident ions is significantly lower than that of a homogenously dense seed coat material as assumed by the computer program. Furthermore, we have found that there exist a number of cracks in the plant seed coat or the embryo membrane, as shown in Fig. 9, particularly under ion bombardment. These cracks may be channels for incident ions to penetrate deeply into the embryo.

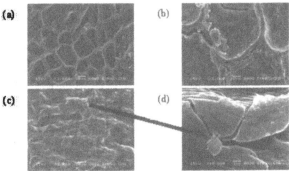

Figure 9. SEM microphotographs of the surfaces of (a) rice seed embryo, (b) flower petunia seed coat, (c) green bean seed embryo, and (d) zoom of the crack from (c).

When ions are implanted into biological matter, depositions of energy and mass and exchanges of momentum and charge combinatively work to produce unexpected effects much more than in condensed matter [1]. The existing and created biological channels can lead to direct interaction of ions with DNA. Ion implantation induced heat, secondary electron and X-ray emissions and produced free radicals as secondary effects may interact with the genetic substance to cause mutation. Some of these have primarily been theoretically discussed [18], but further experimental investigations are needed to provide evidence to support these suppositions. We have carried out some basic studies on this issue.

In order to simulate the final stage of ions interact with DNA, very low-energy and low-fluence nitrogen and argon ions of 1 – 5 keV from with ion beam or plasma with fluences of orders of $10^{11} – 10^{13}$ ions/cm^2 were used to bombard naked plasmid DNA in vacuum. After ion bombardment, gel electrophoresis and DNA sequencing were operated on the DNA. It was found that low-energy low-fluence ions could indeed produce DNA damage in the forms of single strand break, double strand break and multiple double strand breaks to cause mutation. Lighter active nitrogen ions are found more effective in induction of mutation than heavier inert argon ions, probably due to more biological effects. The results confirm that one of the physical mechanisms in ion beam mutation is a small portion of incident ions capable of penetrating the materials covering the nucleus to directly interact with DNA and thus cause mutation [19].

Since at present it is technically difficult in determining the intrinsic features of ion-induced DNA changes, computer simulation becomes a very useful tool to assist in finding answers. Molecular dynamics simulation (MDS) of low-energy ion interaction with DNA has also been involved in the research at CMU [20]. In the MDS, carbon and nitrogen ions at energy of 0.1 – 100 eV bombarded A-DNA (Fig. 10) to simulate the ion interaction with DNA in vacuum. The results show that for the nitrogenous base pairs, poly-AT is more sensitive to argon ion bombardment than poly-GC; for the phosphate group, deoxyribose (sugar) and bases, nitrogen ions interact in the preference sequence with OP, O (in base), O' (in sugar), N, C (in base) and C' (in sugar) atoms and most easily break the O-P bond and followed by C-C, C-C (aromatic) and C-N bonds. The findings demonstrate that low-energy ion beam induced DNA structural modification is not a random but preferential effect.

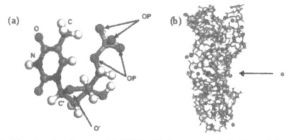

Figure 10. MDS of ion bombardment of A-DNA. (a) Atom type definition of the nucleotide. (b) Ion incidence to DNA.

CONCLUSION

Low-energy ion beam has been demonstrated to have effective multiple effects on induction of mutation and gene transfer due to its multi-factor interaction with biological organisms. The induced mutation is in a broad spectrum and the gene transfer is in a high efficiency. These have brought significant applications to agriculture, horticulture, microbiology and medical science. Relevant mechanisms are being explored. Direct interaction between ions and genetic substance leading to mutation is involved and found that the ion interaction with DNA is preferential in atomic locations and chemical bond types. Electric property changes in the cell envelope induced by ion implantation cause assistance in gene transfer.

ACKNOWLEDGEMENTS

We wish to thank our research staffs for their contributions to the projects, particularly named T. Vilaithing, P. Apavatjrut, A. Krasaechai, B. Phanchaisri, V.S. Lee, P. Nimmanpipug, and R. Chandej. Special thanks are also presented to international collaborators, particularly named Z.L. Yu, I.G. Brown, S. Sangyuenyongpipat, etc. The research has been supported by Thailand Research Fund, National Research Council of Thailand and Thailand Center of Excellence in Physics.

REFERENCES

[1] Yu Zengliang, Eng eds.: Yu Liangdeng, Thiraphat Vilaithong, Ian Brown, *Introduction to Ion Beam Biotechnology* (English Edition), (Springer Science & Business Media, New York, 2006).

[2] L.D. Yu, S. Sangyuenyongpipat, C. Sriprom, C. Thongleurm, C. Tengsirivattana, R. Suwanksum, and T. Vilaithong, *Nucl. Instr. Meth.* **B 257**, 790-795 (2007).

[3] S. Sangyuenyongpipat, L.D. Yu, T. Vilaithong, I.G. Brown, *Nucl. Instr. Meth.* **B 257**, 136-140 (2007).

[4] S. Davydov, L.D. Yu, B. Yotsombat, S. Intarasiri, C. Thongleurm, V. A-no, T. Vilaithong, M.W. Rhodes, *Surf. Coat. Technol.* **131**, 558-562 (2000).

[5] S. Anuntalabhochai, R. Chandej, B. Phanchaisri, L.D. Yu, T. Vilaithong, I.G. Brown, *Appl. Phys. Lett.* **78(16)**, 2393-2395 (2001).

[6] B. Phanchaisri, L.D. Yu, S. Anuntalabhochai, R. Chanej, P. Apavatjrut, T. Vilaithong, I.G. Brown, *Surf. Coat. Technol.* **158-159**, 624-629 (2002).

[7] S. Anuntalabhochai, R. Chandej, M. Sanguansermsri, S. Ladpala , R.W. Cutler and T. Vilaithong, *Surf. Coat. Technol.* **203**, 2521-2524 (2009).

[8] T. Vilaithong, L.D. Yu, P. Apavatjrut, B. Phanchaisri, S. Sangyuenyongpipat, S. Anuntalabhochai, I.G. Brown, *Rad. Phys. Chem.*, **71/3-4**, 927-935 (2004).

[9] S. Sangyuenyongpipat, L.D. Yu, T. Vilaithong, A. Verdaguer, I. Ratera, D.F. Ogletree, O.R. Monteiro and I.G. Brown, *Nucl. Instr. Meth.* **B 227**, 289-298 (2005).

[10] S. Sangyuenyongpipat, L.D. Yu, T. Vilaithong, I.G. Brown, *Nucl. Instr. Meth.* **B 242(1-2)**, 8-11 (2006).

[11] L.D. Yu, T. Vilaithong, B. Phanchaisri, P. Apavatjrut, S. Anuntalabhochai, P. Evans, I.G. Brown, *Nucl. Instr. Meth.* **B 206**, 586-590 (2003).

[12] K. Prakrajang, P. Wanichapichart, S. Anuntalabhochai, L.D. Yu, *Nucl. Instr. Meth.* **B 267**, 1645-1649 (2009).

[13] K. Prakrajang, P. Wanichapichart, D. Suwannakachorn, L.D. Yu, *Thai J. Phys.* (2010), in press.

[14] S. Anuntalabhochai, R. Chandej, B. Phanchaisri, L.D. Yu, S. Promthep, S. Jamjod, and T. Vilaithong, *Proceedings* of the *9th Asia Pacific Physics Conference*, Hanoi, Vietnam, October 25-31, 2004, Session 10: Applied Physics, 10-24C.

[15] B. Phanchaisri, R. Chandet, L.D. Yu, T. Vilaithong, S. Jamjod, S. Anuntalabhochai, *Surf. Coat. Technol.* **201**, 8024-8028 (2007).

[16] A. Krasaechai, L.D. Yu, T. Sirisawad, T. Phornsawatchai, W. Bundithya, U. Taya, S. Anuntalabhochai, T. Vilaithong, *Surf. Coat. Technol.* **203**, 2525-2530 (2009).

[17] S. Mahadtanapuk, M. Sanguansermsri, L.D. Yu, T. Vilaithong and S. Anuntalabhochai, *Surf. Coat. Technol.* **203**, 2546-2549 (2009).

[18] Z.Q. Wei, H.M Xie, G.W. Han, W.J. Li, *Nucl. Inst. and Meth.* **B 95**, 371-378 (1995).

[19] R. Norarat, N. Semsang, S. Anuntalabhochai, and L.D. Yu, *Nucl. Instr. Meth.* **B 267**, 1650-1653 (2009).

[20] C. Ngaojampa, L.D. Yu, T. Vilaithong, P. Nimmanpipug, V.S. Lee, *The Proceedings of the 12th Annual Symposium on Computational Science and Engineering* (ANSCSE 12), 27-29 March 2008, Ubon Rajathanee, Thailand, 127-133.

Ar, 135

biological, 161
biomaterial, 143
biomimetic (chemical reaction), 143
bone, 153

chemical vapor deposition (CVD) (deposition), 27
colloid, 15

electron irradiation, 111

fluidics, 61

ion-beam
    assisted deposition, 69, 93
    processing, 9, 21, 35, 47, 53, 61, 69, 87, 129
ion-implantation, 3, 15, 27, 41, 99, 111, 153, 161
ion-solid interactions, 75, 111, 135, 161

magnetic, 35
    properties, 3
magnetoresistance (magnetic), 35
microelectro-mechanical (MEMS), 61

morphology, 75

nano-indentation, 123
nanostructure, 3, 9, 15, 21, 41, 47, 53, 87, 143, 153
nucleation and growth, 27, 87

Raman spectroscopy, 99, 123
RHEED, 93

scanning
    electron microscopy (SEM), 75
    probe microscopy (SPM), 129
self-assembly, 9
Si, 21
simulation, 69
sputtering, 129, 135

texture, 93
thermoelectric, 41
thermoelectricity, 47
III-V, 53

x-ray
    diffraction (XRD), 99
    photoelectron spectroscopy (XPS), 123

Printed in the United States
By Bookmasters